The Primacy of Stewardship

The Handbook for Christians Who Believe in Democracy

Third Edition: December 2009. ISBN 978-0-9821860-1-5

I

Comments on the first edition of *The Primacy of Stewardship* (2008):

"A beautifully written book. Compelling. Original." JW

"I find it ironic that I could never digest the Bible no matter how I tried. ... The way you interpret it makes sense to me. ... Chapter Four is amazing. I read Chapter Seven ... Your three unforgivable sins at first appear dull and benign. But you show they are indeed our stumbling blocks." HM

"You sure made me think about the teachings of Jesus." TS

"Lucid. I agree with most of what you say ... elegantly written." AC

"You have a wonderful interpretation of *forgiveness*." DG

"Very creative. You have a great argument against war." Methodist Bishop

"You are spot on! You nailed it! Better than best-selling authors, and political pundits. Bravo to you!" JJ

Preface to this Third Edition (Hardcover)

What you are about to read may change your mind.

Jesus Christ (or the source) taught scientifically precise facts about how the universe works. I carefully examine and interpret the detailed content and credibility of the *teachings* as science - not the person or historical issues.

Good stewardship is not simply moral or meritorious behavior that makes one a good or generous person. Good stewardship is scientifically mandatory for a technological animal such as humans on Earth, in order to survive and thrive.

A Christian is obligated to learn about other religions in order to "be reconciled to brothers and sisters" and neighbors near and far.

Individuals and all of human civilization must reconcile science and religion or we will self-destruct. Each chapter of the book could stand alone while they all work together to support the major premises. The purpose of the book is to correct the most important misunderstanding in human history, which is the belief that Jesus was moralizing, talking to Jews about morality and moral behavior and talking about a kingdom of heaven that people go to after they die. The kingdom of heaven that Jesus named is the kingdom of life in the universe. And it has rules.

-- John Manimas, September 2009

Edition History: Each Edition differs in binding.

Third Edition: December 2009. ISBN 978-0-9821860-1-5
Hardcover distributed by Lightning Source, Ingram

Second Edition: June 2009. ISBN: 978-0-9821860-2-2
Softcover distributed by Northshire Bookstore, Manchester, Vermont.

First Edition: December 2008. ISBN: 978-0-9821860-0-8
Georgia Ref on 24lb acid-free Exact Opaque (Ivory), lay-flat binding.
Limited number of copies produced and signed by the author.

The Primacy of Stewardship: Scientific Information in the Gospels

Acknowledgement

I was supported by my mother, her family, and the State of Connecticut. I received scholarships from the Fairfield Lumber Company and Brandeis University. I have always appreciated the living beauty of wood and the fortitude of the Jews (the primary supporters of Brandeis).

Many teachers taught me things I needed to know. Men and women have walked beside me along the way, like protective older siblings. Good friends have given me more than I gave back. The impoverished and the helpless have also been my sponsors. They have made contributions to my cause, educated me, and supported the completion of this project. I am grateful for the recommendations to strengthen the narrative.

I am sorry it took so long. But do not despair -- I say this to myself -- anything can be fixed. Repair is the way of life. To live is to stumble, to be injured, delayed, to get sick. Anything that has lived has had replacement parts, normal wear, a dab of glue, an old borrowed bolt, a spare piece of wire. Whatever works, our job is to step. Every step, any step, if reasonably planned and considered, is counted as a step forward. We sometimes go back but we never go backward. We can never return to exactly what was passed. Whenever we try again to do something right, it is always necessarily and inescapably new. Take heart. Perfect repetition is impossible. We cannot make exactly the same mistake twice. Any mistake that you make today will be refreshingly new and different from all of the exasperating mistakes that were made in the past by your parents and ancestors. There is no need for you to carry the guilt and shame of generations gone. You will have your own inventions, already gathering on your shoulders like the fall of cold, gentle snow.

We continuously tell the same stories to the heavens, stories of how our imaginary super heroes with super powers will stop evil in the universe. But that is not our assignment. The task that has been assigned to us is to stop evil on Earth, with ordinary people who have ordinary powers. In this task, your most important self-deception is the belief that fighting evil and injustice is the work of "leaders." It would be best for you to think of a "leader" as a person who has fallen into a deep well. You call down to them, and you become aware of splashing sounds. You call down again, and hear your echo asking them "What's happening?" And the leader responds, often with great, profound wisdom, and heartfelt encouragement. Leaders are positive by nature. They equate death with re-birth, destruction with opportunity and progress. Many fantastically beautiful speeches have been composed and delivered over the centuries, perhaps thousands. I would like to show all of them here, to sustain your heart and uplift your spirit. However, in the interest of saving time and paper, I will summarize all of the historic speeches from the great minds of the inspiring leaders who have blessed our human civilization with their expert advice into one concise speech: "Don't worry. Everything is fine."

Foreword

The theme of this book is that *Social Justice is the Gospel*. This is scientific information, not just a message about morality or how to be nice to your neighbors. I do not claim to have a *personal* relationship with Jesus. I do feel that I have a *professional* relationship with Jesus. I am in the business of correcting errors and revealing the truth. I recognize the social worker in Christ. He empowered people with information. *The Primacy of Stewardship* sheds light on that vital information. I claim no special role, status, or place. I am only doing my job. Perhaps I ate some fish and bread a long time ago, and I have a good memory. Perhaps I saw him and heard him and felt his breath on my face when he spoke to me and said, "Ask, and you shall receive. Knock, and it shall be opened to you." (Matthew 7: 7 -11).

My work, presented here, can and will be described as an "interpretation" of the Gospels. That is fine with me, but it is important that I openly acknowledge that I am certain that I have understood the message of Jesus Christ with precision of the highest quality. I do not apologize for my interpretation, which is different from the interpretation of biblical scholars, religious teachers, priests, ministers, bishops, the Pope and experts in history and anthropology and ethics and moral philosophy. I disagree with the proposition, openly advocated by Joseph Campbell and implied by Carl Jung, that Jesus was addressing the "inner life" or the "collective unconscious" of human society, or any form of morality that is deemed by the divided human brain to be in some way separate or distinct from practicality. The message of Jesus is not only scientific, it is profoundly scientific; it is the most important scientific information we have ever received and have in our possession. Matthew Chapter 23 is a description of who and what we are, a species of technological animal that has a persistent problem with authority, sexuality and social status, and the essential capacity to change one's mind when new evidence tells us that old ideas are errors in understanding. This pattern of human "authoritarianism" is everywhere, in families, in social groups, in tribes, neighborhoods, institutions, clubs, towns, cities, states, nations. It is as much a part of us as our human scent that is always with us when we walk through a field or forest. Wherever we are, there is the smell of authoritarianism. Our obsession with status, conformity, power and hierarchy is as much a part of us as our skin. It is stamped on the human identity. We can understand it and resist it. We can mediate its negative effects with our rational mind, with equality, democracy, and with social and economic justice. We can correct our innate faults with *good stewardship*. The Good Shepherd is not a *nice man*. The Good Shepherd is the species of intelligent being that survives and thrives in a universe governed by physical and spiritual laws, a universe that has no regrets, only survivors.

Jesus never said, "Take me to your leader," or "I have come to talk to your philosophers, your kings and warriors, your scholars, the rich and powerful." He sat at the table with "sinners" and said he had been sent to speak to the "sinners" and the poor and the powerless. When his disciples tried to chase children away, not only did Jesus tell his disciples to let the children come to him, he also informed his disciples (Matthew 19: 13 -15) that they would not be admitted into the kingdom of heaven unless they were humble and awed by the universe like little children.

The authority most worthy of one's respect is the authority of truth, not the authority of "office" or social status or of certificates and diplomas. Academic degrees may become a weapon of mass destruction if their owners have no ethics and teach doctrines of conformity and dogmas of compliance or deform the truth for a price. Jesus speaks directly to you and to me, directly to the lonely, the dejected, the rejected, the confused, the uncertain, the afraid, the average reasonable person. You are not enlightened by the hierarchy. You are enlightened from that moment you believe you are fully qualified to hear the message of Jesus directly within the record of his words and actions. He spoke to the crowds, to the poor and the sick who wanted truth, reality, healing. He brought all three to us, to the people. "Hear, all those who have ears to hear. See, all those who have eyes to see!" he said. I want everyone to see what Jesus really taught, so that no one will be fooled by those who use religion as a confidence game, the liars and hypocrites, the authoritarians. My viewpoint is that the most important truths are the points where science and religion converge. I present the teaching of Jesus as a unified whole, a message to humanity of supreme importance that is consistent with biological and social science. I stand on common ground with everyone who believes the message of the Gospels is important. However, I stand on new ground with my viewpoint that we have disastrously underestimated the scientific content of Christ's teaching, the reality that the Good Steward is the highest level of evolutionary development for any intelligent being, on any planet, in this one universe. By "scientific content" I mean that the Gospels convey information about the physical reality of the universe, not about a separate non-physical or mythical or spiritual realm.

Biblical citations are from a Roman Catholic version I received in 1958:
Holy Bible. O. T. Confraternity-Douay Version; N. T. Confraternity Edition, Confraternity of Christian Doctrine. New York: Catholic Book Publishing Company, 1957. I believe this translation was the most accurate ever written from the viewpoint of linguistics, and will remain so until we obtain more information about ancient languages. This evaluation is still subject to the criticism that the church authoritarians removed extremely important information that did not support their political and doctrinal goals in the fourth century. At age fourteen, I told Father Dunn of the Church of the Holy Family that I wanted to read the Bible before making a personal decision as to whether I would pursue Holy Orders (become a priest). I asked the cost. He said six dollars (it may have been three dollars). I earned the money and returned. When I offered the payment, he handed me the Bible and said,

"The Bible is free." I knew instantly that he had only tested me so that I (and he) would know the measure of my interest. I also remember distinctly leaving the rectory that day, with my Bible in hand, believing that the truth in the Bible was for me priceless and therefore free. Many years later I inquired as to the welfare of Father Dunn. I was told that he had moved to another state and married a nun. Upon hearing of his life's path, I believed that I got my Bible from the right man. I did not become a priest because after reading the Bible I concluded that Jesus recommended the practice of morality in the marketplace, not behind the walls of a fortress. Since that time, we have been painfully disappointed by the disclosure of the failures of priests to live up to the rules that they make for us and for themselves. Our spiritual and political authorities persistently fail to make good decisions. This presents us with the threatening reality that they are failures as authorities and we are failures at selecting our authorities and raising people to positions of power. There is only one way to cure this societal sickness: understand and practice Good Stewardship. The Introduction that follows this Foreword shows that one priest, the Reverend Anthony de Mello, agrees that Jesus speaks to us directly, and that if there is an "expert" standing between you and Jesus Christ, you are wise to tell that expert to step aside. It is rude to stand between Jesus and the person he is speaking to.

John Manimas, 2008

Introduction

The Gospels are intended to be heard by the average person directly from Jesus and the evangelists (Matthew, Mark, Luke and John). No one should be lazy by relying on the professionals and experts to tell us the truth. Many of those with institutional power believe that their duty is to protect you from the truth or interpret the truth for you, or to hoard it, not publish it for everyone to see. I invite you to enjoy two selections from *The Song of the Bird*, a book of wisdom by Anthony de Mello. Reverend de Mello was a Jesuit priest who lived in India. Reverend de Mello's writings tell us that being religious is not comprised entirely of social conformity.

The Professionals and *The Experts*, cited from *The Song of the Bird* (Image Books, Doubleday, 1982), p. 49-50:

> The Professionals
> My religious life has been taken over by professionals. To
> learn to pray I need a spiritual director; to discover God's will
> for me I consult an expert in discernment; to understand the
> Bible I consult a scripture scholar; to know if I have sinned
> or not I need the moral theologian; and to have my sins
> forgiven I kneel before the priest.
>
> *A native king in the South Seas Islands*
> *was giving a banquet in honor of*
> *a distinguished guest from the West.*
>
> *When the time came to praise the guest, His*
> *Majesty remained seated on the floor*
> *while a professional orator,*
> *engaged for the occasion,*
> *eulogized the visitor.*
>
> *After the panegyric, the guest*
> *rose to speak.*
> *His Majesty gently held him back.*
> *"Don't stand up," he said. "I have engaged*
> *an orator for you too. In our island*
> *we don't leave public speaking to amateurs."*

I wonder, would God appreciate it if I became more amateur in my relationship with him?

The Experts, a Sufi Tale:

A dead man suddenly came to life
and began to pound on the lid of the coffin.

The lid was raised; the man
sat up. "What are you doing?"
he said to the assembled crowd.
"I am not dead."

His words were met with silent disbelief.
Finally one of the mourners said,
"Friend, both the doctors and the priests
have certified that you are dead.
So dead you are."
And he was duly buried.

****

I (John Manimas) have written this book to be clear about what I believe and so that my neighbors will understand my God and my religion. According to Jesus, it is God's desire that we be reconciled with one another, that any dispute or conflict between us be settled before we each bring our gifts to the altar of God (Matthew 5: 21-26). How will you be reconciled to me, and I to you, except that we each know what the other believes?

Part I: The Universal Jesus

Chapter One: My God and Your God

<u>Jesus' relationship to God</u>:
I believe in God, in democracy, in equality, and in the primacy of stewardship. I believe that much of Jesus' teaching is pertinent to our concept of democracy, and that Jesus taught us that serious democracy requires serious equality. Jesus referred to himself as the "Son of Man." He is alleged to have said that he was a son of God, but he also stated that all humans were the children of God. Therefore, he did not identify himself as possessing a divinity that was different from the holiness in the life and soul of every human being.

<u>Jesus' Teaching and Example</u>:
Jesus taught that good stewardship was of primary importance, and that good behavior for any human being always meant nurturing life when vulnerable or weak and liberating life when strong, productive and independent or inspired. Ritual and appearances are secondary. Those who are poor and who appear to fail in life have most likely been subject to circumstances they could not control. They may have been injured or hindered by the unfairness of human hierarchies or the selfishness of the strong who sometimes act as though they have a right to exploit and manipulate those who are weaker or less able. They may have been neglected, mistreated or even tortured as children, which would have directed their personal development toward using any means, moral or not, to survive in a dangerous, unjust, and lonely environment. A mistreated child is likely to grow up to be an adult who expects others to lie, cheat and steal, and to hurt them for gain or entertainment. Some adults become criminals because they see the world as very predictable. What is predictable for them is that they will not be treated fairly no matter how good they behave, and they will always be blamed for whatever goes wrong, whether

1

they are responsible or not. Their crimes are their revenge, their vented anger, against a human society that they know as incapable of justice.

Forgiveness for Commitment:
Jesus advises us to forgive offenses, but not lightly. Whenever Jesus was met by a person who said, "I have sinned. What shall I do?" he responded by advising them to shed any burdensome material possessions and "If you want forgiveness, go forward and sin no more." This expectation that the offense will never be committed again is not "permissiveness" and neither is it tolerance of evil. It is in fact the most serious commitment and most difficult sentence that can be imposed on anyone for any offense. What more could be demanded of a person than that they never commit such an offense again? Never lie again. Never steal again. Never lash out in anger again. Never misbehave sexually again. Never cheat again. Never again commit a lie of omission by not telling someone what they need to know and have a right to know. Jesus was not soft. Basically, he said that your forgiveness is conditional. If you really want to be forgiven, if you want to rejoin human society rather than being shunned or banished, renewal of your membership in the human community is dependent on your genuine vow to never commit the same offence again. This is in fact a scientific approach, because other than "restitution" what we want from anyone who has engaged in destructive behavior is the promise to not repeat the offense.

Human Worthiness as a Scientific Message:
All of Jesus' teaching is scientific. His teaching included physical science, which has been forwarded to us as narrative descriptions of "miracles." Much of his teaching was social science, including psychology, sociology and political science, but much of that was taken by his listeners to be "moral" lessons to be used to define "sin." Jesus was not concerned with sin and punishment, but the people were then, as many people are today. People try to construct a religion on a foundation of a handful of "sins," especially

sexual offenses, and what they produce bears no resemblance to the clear and elegant message of Jesus. Jesus was a missionary. His main message was not that he was God, but that we were worthy of our place on Earth. The point of his mission was to let us know that we had been judged capable of being good stewards, or good shepherds, and that we would have a chance to prove that judgment correct. He came to us not to tell us who he was, but who and what we are, and what we need to know in order to survive and thrive on this planet and in this universe. We are technological animals who can cause great destruction, but we also can grow and learn from our experience and become the caretakers of life on Earth. As for life beyond the planet Earth, we might find that taking care of life on Earth is enough when we come to an understanding of all the responsibilities that assignment entails.

Shall We Compare Gods?

In the film entitled "Revenge of the Pink Panther" (1978), there is a scene where the comical Inspector Clouseau (actor Peter Sellers) and his attractive female friend Simone (actress Dyan Cannon) enter her apartment for the first time. The bumbling Clouseau has just accidentally saved Simone from hired killers. Clouseau and Simone are soaking wet after running to her apartment in the rain.

> Simone exclaims, "Oh my God!"
> Inspector Clouseau responds, "Yes, mine too."

This is surprisingly funny and at the same time it conveys a profoundly wise observation about human nature. We each have our own God in the sense that we each have our own understanding of who or what God is. Or we have a hazy definition that makes God a "mystery" and free from any of the usual characteristics of a real entity in the real, physical universe. It is as though "God" is just an all purpose invention to take responsibility for us whenever we do not understand what is happening or why it is happening. After we have grown up and our parents are no longer in charge of us, and no longer in charge of rewards and punishments, we may appeal to God to punish our siblings and our friends and neighbors whenever we feel that they are behaving badly or not treating us the way we ought to be treated.

When someone says, "I believe in God," we really know nothing about their true moral standards or their actual moral behavior in challenging situations. Yet the world is full of people who talk and act as though the fact that they say "I believe in God" makes them religious and is evidence, in itself, that they have faith and have some form of "protection" from a God whose role is apparently to supervise the universe and make sure everyone, like children in a nursery, gets appropriately punished and rewarded. This "religion" found in Judaism, Christianity and Islam, implies that human beings would not do the right thing unless they are constantly reminded of the threat of horrible punishment and the enticement of an awesome, and eternal, reward. To me, such religious institutions actually destroy real faith, because they promote the doctrine that human beings are naturally evil. That is not what Jesus conveyed. According to the Gospels, he came to us on a mission of mercy and encouragement, to tell us that we can be worthy of the kingdom of heaven. He came to us even though he knew that the authorities on Earth would probably resent him and punish him for claiming to have the moral authority they want for themselves. Therefore, Jesus came to us under threat of death, to tell us what we needed to know and to tell us that we are worthy enough to become good stewards and become permanent members of the kingdom of heaven. His mission is an act of faith *in us*. His mission and his life and teachings convey the message that we are good, not that we are bad and will only do good with a sword hanging over our neck. There is a science of "behaviorism" and of psychology that encompasses our best efforts to study human motivation and human behavior. To me, these social sciences are the human act of real faith, the faith that we can understand the Creation

4

and the Creator by doing the work of the heart, mind and soul, which means searching for the truth that is within our capacity to understand. This is the capacity to understand that was given to us, if one is a fundamentalist, by the Creator God. If one argues that the capacity to understand that was given to us by God is insufficient for us to understand the Creation and the Creator, then the defective Creation discredits the Creator.

The Authoritarian Creator God:

The Creator God as defined by fundamentalist and some conservative Jews, Christians and Muslims is an authoritarian and punitive God by definition. First, this God is a male only, who has infinite power and control over the universe. He can even suspend the laws of physics, or change them if he chooses. Since He created everything, everything is under his control. Yet, these same people, including the Roman Catholic Pope, insist that we have free will and must be held responsible for our actions. That is why we can commit a sin and be punished by God. Rational people throughout the world, including those who are affiliated with a religious institution and those who are not so affiliated, wonder why a God who has total power over the universe would invent and create defective humans. It makes no sense. When we manufacture things, we make every effort to have "zero defects." If we build a million bicycles, we know that a few will have come out wrong, and will need to be repaired or replaced. However, the percentage of "defective" products is expected to be very low, perhaps one percent, or even less. How could a powerful and competent God create a type of human being who invests tremendous time, effort and resources in the goal of killing his neighbors? Why create humans with sexual drive but then make sexual activity a sin? The problems of this all-powerful and controlling God go on and on. It makes no sense. But, those who claim to believe in this Creator God have cornered the market on God products. They insist that if we do not believe in *their Creator God* we are a condemned sinner or an atheist or an *infidel* (one who has no faith). If we do not believe in *their God*, we cannot

5

be religious, cannot please God, cannot be protected by God, and we surely are condemned to go to Hell forever. In fact, since it is certain that God is going to punish us, it is their duty to help God by taking on the task of punishing us right now, here on Earth, for not complying with their concept of God. This is obviously contradictory because one must wonder why an all-powerful Creator of the Universe needs help from humans. There could be a God other than this Authoritarian God. There could be a God who has many good qualities, but who came into existence after the Universe was created by Nature. Such a God that is a product of Nature, the same as the stars and planets and plants and animals and we humans, who I see as technological animals, could have many qualities similar to the qualities of a Creator God, such as being concerned with our welfare, and wanting us to survive and thrive and become the best that we can be, but is not a "person" who existed before the universe existed. Such a "non-creator" God could be subject to the laws of Nature just as we are, rather than the bizarre Creator God who plays the universe like a giant video game.

The Gods of Men -- and Women:
The other amazingly obvious problem with the male Creator God is precisely that He is Male and his holiness and sanctity and power and spirituality and control over the universe is reserved for Men Only, kind of like having a moustache or urinating while standing. Don't people ever wonder why, if a woman was good enough to be the "mother of God," she's not good enough to be a priest? Not good enough to be vested with authority? Not good enough to be a parent of the species? Why, when in fact women are the primary parents of children and have been for thousands of years, do we make our Parent in Heaven a man? If a woman is the right choice to take care of every child born to humankind, son or daughter, why do the institutionalized religions make our spiritual parent a man? If we are going to avoid the authoritarian God, and the authoritarian defining of God, we absolutely must include the freedom for an individual or group to define God as female, or as

a Mother, or as a creative female spirit, or as a Holy Spirit that is female or both male and female. The institutions of religion are obviously "homo-gender" and treat the female side of the human personality as though it is inherently defective, or sinful, or dangerous or otherwise inferior and subordinate. Go back to the Bible. Genesis. Eve was made from Adam's rib. That means that genetically Adam and Eve are the same person.

Religion May Sustain a Lack of Real Faith:
Ask this question: "What is the most fundamental principle of religion?" Then ask: "What is the most fundamental principle of science?"

My answer would be the same for both questions. The most fundamental principle for both religion and science is: We can understand the universe as it really is or is evolving or becoming. This is real faith. This is what has to be believed first, before anything else is worth the effort. If one does not believe that we can understand the universe as it really is, then both religion and science are a waste of time. If one does believe that we can understand the universe as it really is, then that person will stop blaming God for what happens. That person will not believe our destiny is all some kind of intractable, irrational mystery that only God knows or can control. Either we can understand reality or we cannot. And if one insists that a Creator God created the universe, then that same Creator God created a foolish humankind who cannot understand what God created and cannot understand The Creator. Real faith is destroyed by the idea that we cannot understand reality but must believe in mysteries and a God who has magical powers and who punishes us for having defects that He created.

For me, Nature is the Creator. I am certain that Hell does not exist, because I know Nature. Nature is economical. Nature does not waste any effort or energy, and does not create anything that is unnecessary to sustain life. There is no Hell because Hell would require a great quantity of matter and energy to be sustained. Hell would be a waste of matter and energy for no productive purpose that is related to life. It would consume an enormous quantity of energy, but would serve no useful purpose. Inflicting unnecessary pain is not something a good God would do. The consequences of doing the wrong thing come to us, with absolute certainty, from Nature. No magical manager of the universe is needed to assure that the laws of Nature shall remain in effect. In our search for understanding, the reason we ask "Why?" is because there is an answer, an answer that we can understand. For every question that we are smart enough to ask, we are smart enough to find and understand the answer. This is real faith, not just faith in ourselves but faith in our Creator, whomever or whatever creator that may be. Jesus tells us that God will surely know what we each really believe, not by what we say but by the fruits or results of our actions (Matthew 12: 31-37). What are the results of our actions? Do we add to human understanding, or do we propagate fear of mysterious forces? Which of these two is a true act of faith?

A God Who Empowers Us With Information:
My God sent Jesus to us to tell us what we need to know, to tell us that good stewardship is our primary obligation and our primary goal. The reason we need to be good stewards is because any technological animal that has the capacity to change and control natural processes must be a good caretaker of the life-supporting environment. Otherwise, that creature will self-destruct.

8

Chapter Two: The Tree of Life: Jesus Taught Evolution

In Matthew Chapter 13, Jesus tells the parable of the mustard seed, which is unlike any other in that it makes no sense and implies, if taken as recorded, that either Jesus was a fool who knew nothing about trees and birds or the Gospel writer completely misunderstood what Jesus said. Or, there is a third possibility, which I believe is what occurred. That third explanation is that the original parable was misunderstood by its listeners, and the misunderstanding was preserved in all following translations. Hear again, the parable of the mustard seed, and look at what appears to be a strangely ignorant statement about a tree and about birds. A metaphor has to be grounded in reality. A comparison has to have some basis in reality or it does not serve the purpose of clarity and understanding. Strictly speaking, when using "like" or "as" we have what is called a "simile." A viable metaphor is to say that a fish swims through water like a bird flies through the air. And an even better metaphor is to say that the sting ray flies through water like a bird flies through the air. The sting ray swimming even looks like a bird flying because of the undulation of its wing-shaped "arms." However, it is a junk metaphor to say that a dog runs on land like a fish swims through water. Look at the metaphor in the accepted wording of the parable of the mustard seed. I see it as a "junk" metaphor that discredits Jesus' intelligence. I say that if we changed just two words that would present the ancient listeners with the concept of evolution, a concept that they could not grasp, we would then have the true original parable. Here is the parable as commonly interpreted in the Bible:

> The Parable of the Mustard Seed: 31. Another parable he set
> before them, saying, "The kingdom of heaven is like a grain
> of mustard seed, which a man took and sowed in his field.
> 32. This indeed is the smallest of all the seeds; but when it
> grows up it is larger than any herb and becomes a tree, so

that the birds of the air come and dwell in its branches."

If we accept this wording as the accurate original, we have three challenging issues that suggest the parable cannot be from the mouth of a man with knowledge.

First: The mustard seed is not the smallest of all seeds;
Second: The mustard seed does not become a tree;
Third: The birds of the air do not all dwell in the branches of any single type of tree. Each species of bird chooses certain types of trees. We are not likely to see an eagle and a sparrow in the same tree, nor a crow and a finch.

In Mark, the parable states that the mustard tree "puts out great branches, so that the birds of the air can dwell beneath its shade." This still raises questions about birds dwelling beneath the shade of a tree's branches. Most birds "of the air" dwell in the branches themselves, and in nests.

And this is what is stated in Luke: "and the birds of the air dwelt in its branches."

When I studied the Four Gospels, re-reading them three times and taking notes, I could not help but focus on Jesus' frequent references to "the kingdom of heaven." A careful study of the Gospel naturally provokes this question. What is the kingdom of heaven? I concluded, based on all the parables where the kingdom of heaven is "likened" to something familiar on Earth, that the kingdom of heaven is the kingdom of life. In fact, I concluded that the kingdom of heaven means the kingdom of life in the universe, not just on Earth. In several parables, Jesus said that the kingdom of heaven "is like a farmer." Why would he persistently compare the kingdom of life to a farmer? Jesus also persistently told parables about "good servants" and "bad servants," or about good stewards and bad stewards, about good shepherds and about masters or landowners or householders or farmers or vineyard owners who are responsible for the productivity of the enterprise entrusted to them and for the people who do the work for that enterprise. He consistently presented the stories of "good stewards" as people who take care of the people and property in their care, so that they can be appropriately productive for the benefit of all concerned, for the benefit of the living human community. This is the theme that I perceive as being at the core of Jesus' message of social justice and spiritual competence. Social justice and spiritual competence arise out of good stewardship, and out of proper care for life and the property of the community that supports the life of the community. The property itself is not to be worshiped, and no one should allow himself to become attached to property as though it were more important than people. However, productive property that produces the food, clothing, housing and tools that people need also needs to be taken care of in order to nurture and liberate life.

Since I see stewardship as the core principle being taught, and because I see the persistent implication that the kingdom of heaven is the kingdom of life in the universe, I concluded that the parable of the mustard seed had to be consistent with the principle of stewardship and with the concept that the kingdom of heaven is the kingdom of life. Therefore, in the parable of the mustard seed, we have to overlook the apparent problems with the statement that the mustard seed is the smallest of all seeds, and that it becomes a tree and that all or many types of birds dwell in or under its branches. This parable does not make any meaningful sense unless Jesus is saying something about the kingdom of life in the universe. That is how the parable begins: "The kingdom of heaven is like... ." Now I take the version I have from Matthew and I delete [the birds of the air come and dwell] and insert (*all the little creatures*):

11

31. Another parable he set before them, saying, "The kingdom of heaven is like a grain of mustard seed, which a man took and sowed in his field. 32. This indeed is the smallest of all the seeds; but when it grows up it is larger than any herb and becomes a tree, so that *all the little creatures are* in its branches." Or, "*...are its branches.*" (Words in italics in this passage are the author's interpretation.)

I believe that this is the original parable, and the ancient disciples and other listeners, having no idea what Jesus was talking about, heard "all the little creatures" and were certain that Jesus must have been talking about birds nesting or perching in trees. I ask the reader to think about what this parable must mean, if one agrees that the kingdom of heaven is the kingdom of life, and the primary principle is the practice of good stewardship. The parable, with either wording, has to be saying that the kingdom of heaven is the framework for living creatures to *live in*. This parable says that the *tree of life* that supports small, vulnerable creatures (such as birds), starts as a tiny seed, but grows to have many branches, and the creatures dwell in its branches. If we claim that the reference to "birds" is accurate, then the parable is comparing the kingdom of life to a kind of shelter, a living plant that provides a home for birds. This metaphor is more or less acceptable, but it is rather specific, and limiting. This interpretation suggests that the kingdom of life provides a home for the kingdom of life. This is true enough, but if the tree is actually the kingdom of life in the heavens, then it must be the *tree of all life*, and that would make it *The Tree of Life*, a religious symbol for life in the universe that is found throughout the world and was recognized among our oldest civilizations long before Jesus spoke to us. This is enough to persuade me that the "mustard tree" is a mistranslation or double or triple mistranslation. Jesus said that the kingdom of heaven, being the kingdom of life, is like a tree that begins as a small seed but then generates many branches and becomes the home of many plants and animals that are "in its branches" and in fact *are its branches*, the branches of life forms that is familiar to us as the tree-shaped diagram that depicts the evolution of different species from the smallest ancient origin. Jesus' "mustard tree" is the Tree of Life and the Tree of Evolution.

Chapter Three: Panspermia
(Life is disbursed throughout the universe)

Jesus repeatedly likened the kingdom of heaven to a farmer. He also said that the kingdom of heaven was like a householder, which person was in his time, as well as now, a woman. A householder is the person who takes care of all of the affairs of a household, and in some cases, a household is large and involves diverse property of great value.

Jesus' description of the kingdom of heaven as a sower planting seeds randomly is of the greatest interest, because this is not ordinary behavior for any farmer, in the past or present, and Jesus does say that this farmer *is the kingdom of heaven*. This parable and others led me to the conclusion that the *kingdom of heaven* is *the kingdom of life in the universe*. Look at the parable, and how the farmer is casting seed everywhere, on poor soil as well as good soil. Do you know any farmers who do that? Not likely. Farmers on Earth, who want the best crop for their labor, plant seeds very deliberately in the best soil available to them, and only on that ground that they will have the right to cultivate, irrigate, nurture and harvest the crop from, the fruits of their labor. They don't throw any seed "by the wayside."

> Matthew 13: 3-9: 3. And he spoke to them many things in parables, saying, "Behold, the sower went out to sow. 4. And as he sowed, some seeds fell by the wayside, and the birds came and ate them up. 5. And other seeds fell upon rocky ground, where they had not much earth; and they sprang up at once because they had no depth of earth; 6. but when the sun rose they were scorched, and because they had no root they withered away. 7. And other seeds fell among thorns; and the thorns grew up and choked them. 8. And other seeds fell upon good ground and yielded fruit, some a

hundredfold, some sixtyfold, and some thirtyfold. 9. He who
has ears to hear, let him hear."

Now this "parable" is extremely important for three reasons. One, it is
presented to us, that is, passed down to us, as a quote of Jesus. It is not given
to us as an interpretation from one of his disciples. The quote means "Jesus
said this."

Two, the farmer is either careless and incompetent or the farmer is different
in some significant way from farmers on Earth. Most farmers on Earth,
especially the farmers that would have heard Jesus' parable in his lifetime,
were not rich. Seed is precious. Ordinarily, farmers do not plant seed
carelessly. No seed would be "falling by the wayside," or dropped on stones
or in a patch of thorns. Therefore, my scientific analysis of this parable is
that the kingdom of life in the universe is either Nature alone or Nature with
the help of intelligent beings who participate in the process of "throwing"
seed (dry dormant spores) into the cosmos, randomly in all directions, for the
purpose of having new life forms evolve on planets in accordance with each
planet's different conditions.

Three, Jesus concludes the metaphorical parable by saying "He who has ears
to hear, let him hear." The phrase "He who has ears to hear" is of course not
a reference to the ears on the side of one's head, but rather to the "ears" in
one's brain. This phrase really means "He who is able to understand this, let
him understand accordingly." In other words, some people may not be able
to cope with the concept that life pervades the universe, and that life came to
Earth from the cosmos, or space, or possibly just "fell to Earth" naturally
because life spores float in space randomly until they come to rest in the
atmosphere or in the sea or on the land of a life-supporting planet. They are
destroyed or damaged if they encounter the killing heat of a star or some
other destructive force or event.

In 1984, scientist Fred Hoyle described in *The Intelligent Universe* (New York: Holt, Rinehart and Winston) how early space probes found spores in space. They were not exactly like what we have here on our own Earth, but they were like bacterial spores, with very protective shells, their natural "space capsules" that enabled them to travel through the vacuum and extreme temperatures of space. How could a spore remain alive in space? First, being in a vacuum is not a problem. We preserve foods on Earth by having them "vacuum packed." Extreme cold is not a problem for anything that is extremely dry. Freezing damages living tissue only when it is fully "hydrated," that is, full of water. If a spore in space is a special kind of seed that is exceptionally dry, the cold will not harm it. Extreme heat could harm it, but it does not encounter extreme heat unless it collides with a star. That would be "scorching" ground. All other options, however, a cold, dry planet; a warm, wet planet; a large planet, a small planet; dark or bright; could all allow the spore to "sprout" and become a life form that can survive and reproduce and evolve in its new home. This could happen as a process of Nature alone, or with the aid of an intelligent being type of "farmer." Either way, the metaphor holds true, and is sound science, if dormant life spores travel through space until they are captured by the gravity of a planet. Jesus taught science. Stop accepting the nonsense that Jesus is a friendly police officer rescuing a kitten from a tree. Stop accepting the nonsense that access to the kingdom of heaven is yours so long as you *love Jesus*, or love your neighbor, or have a warm heart. Jesus said: Let them hear (understand) who have ears to hear (have the brains to understand). This means, whether we like it or not, Jesus was teaching something that some people can understand and others might not be able to understand. This means that we need knowledge, intellect, brain power, to understand the kingdom of heaven, to understand what the kingdom of heaven is and who Jesus is and what he is really teaching. This is very different from the line of the churches and moral literati who keep telling you that "all you need is love" and a hug

17

from Jesus. We each need understanding in order to know who and what we really are; who and what the kingdom of heaven is; who and what Jesus is.

Jesus is a soldier on a mission, a kind of commander. His own disciples do not usually appear to understand this, but Jesus' serious authority was recognized by an officer (centurion) of the Roman legion.
Look at the Gospel story in Matthew 8: 5-13.

> 5. Now when he had entered Capharnaum, there came to him a centurion who entreated him, 6. saying, "Lord, my servant is lying sick in the house, paralyzed, and is grievously afflicted." 7. Jesus said, "I will come and cure him." 8. But in answer the centurion said, "Lord, I am not worthy that thou shouldst come under my roof, but only say the word, and my servant will be healed. 9. For I too am a man subject to authority, and have soldiers subject to me; and I say to one, 'Go,' and he goes; and to another, 'Come,' and he comes; and to my servant, 'Do this,' and he does it."
>
> 10. And when Jesus heard this, he marveled, and said to those who were following him, "Amen I say to you, I have not found such great faith in Israel. 11. And I tell you that many will come from the east and from the west, and will feast with Abraham and Isaac and Jacob in the kingdom of heaven, 12. but the children of the kingdom will be put forth into the darkness outside; there will be the weeping and the gnashing of teeth." 13. Then Jesus said to the centurion, "Go thy way; as thou hast believed, so be it done to thee." And the servant was healed in that hour.

This is an incredibly blatant clue in the Gospel to who Jesus really is, but it is glossed over, twisted and interpreted to nothing by most Christian clergy and

Christian philosophers. They usually present this to a congregation, or the general public, as a moralistic metaphor. They say that the "servant" of the centurion is actually a symbol for the centurion's soul, or the soul of any individual. Then, if you believe, your soul will be "healed," or "saved." They argue that this story is another parable that is not about the real, physical world but about the "inner life," of the individual, the moral life. I hate this "moralizing" with a passion. It makes me want to rip paper, or grind boards into sawdust. What really happened is a human military officer recognized the honorable power and authority of another officer, even though that officer being Jesus did not receive his commission from only an Earthly government. This is real, as real as swords and fireballs and rivers of blood. There are two astoundingly important concepts fired at us here, like shrapnel from the mouth of a revolutionary cannon, and they are ignored or turned to mashed potatoes by the church.

The Centurion, Jesus, and the respected servant of equal value:
One, Jesus *marveled*, and affirmed in no uncertain terms that the centurion is correct. Jesus labeled the centurion's understanding of who and what he (Jesus) is as "*faith*," as the belief in a truth that escapes everyone around him, in all of Israel, including his disciples. Thus, the *faith* of the centurion is sound knowledge, not blind belief in something illogical. Two, a factor that I have never heard mentioned in any sermon or exegesis of this passage, is the genuine depth of love, concern and passionate equality of worth that the centurion feels toward his servant. In Jesus' time, as well as our own, many would see the servant as "less important," not worthy of any special attention, a lesser being of lesser worth. The centurion speaks of his servant as though he (or she - the gender is not truly established - Jesus assumed a "him" as would most people.) is a spouse or a child. This centurion, barely seen, unnamed, speaking only a few words, could be the subject of a heavy novel. He is a man of war who feels for the health and welfare of his servant. He is like Jesus. Jesus is like him. That's why he knew he could ask Jesus to

care about his servant only as much as he does. This is Jesus, the commander who has come to teach, and who cares about everyone equally, the high and the low, the officer and the servant, the priests and the "sinners." Do you want to know who Jesus really is? Ask this Roman centurion. He knows.

Chapter Four: Stewardship is The Gospel

The primacy of stewardship means that stewardship, being a good shepherd, taking care of the people and property within one's reach, is the primary theme and primary lesson in Jesus' teaching. It is more important than anything else. The reason it is more important than anything else is not because it is morally right, or because Jesus wants us to be responsible citizens or responsible neighbors. It is of the greatest importance because it is the law of the universe. The primacy of stewardship is a scientific principal that governs the evolutionary path of any and every technological animal and intelligent species.

From my viewpoint, all of the Biblical scholars and philosophers and moralizers over the centuries who argued that Jesus was a teacher of morality or of "the inner life" have made a tragic error. It has been very destructive for the human species to go on for so long thinking that Jesus was a "nice man" who taught us to rescue kittens from trees and pat children on the head. It is just as wrong to raise the seriousness of such moral teaching to responsible adult behavior in order for the individual to be "righteous" and polite and have moral "integrity." Even if we raise the meaning of moral behavior to a higher level of personal responsibility and character, we are still treating Jesus' teaching as being about "morality," and about being good, especially about being good and not getting any positive consequences for being good on the Earth. The reason we cannot expect positive consequences on Earth for being good is because the world is "not fair" and the social and physical laws of the universe do not necessarily reward good behavior. We are all taught, no matter how much morality is praised, that the practical outcomes of human behavior and the way the world works gives power to the strong, the aggressive, the violent, the ambitious, the selfish, the powerful, the rich, those who take what they want. In short, the purpose of morality is to be able to respect oneself and to receive some intangible rewards either here on Earth or later in an imagined but separate existence called "heaven."

This view of Jesus' teaching enables people throughout the world to dismiss such ideas as childish, impractical and nothing more than a scheme to make people submissive and compliant. My viewpoint changes all that, because for me good stewardship is the physical law of the real, physical universe. It is the state of any technological animal that has evolved to the highest form of life achievable. I say that the "good steward" is in fact the highest stage of the evolution of life. When a technological animal such as earthly human beings achieve the true status of "good steward," we will then have completed a full cycle of evolutionary development. We will be life taking care of itself. We are already life taking care of itself, but when we achieve the true status of good stewardship, we will be *competent* at taking care of life, which means taking care of ourselves by taking care of all of life in the process and always in a competent manner with no *unintended consequences.*

Jesus spoke of the central importance of good stewardship in more than a dozen parables. His references to good stewardship appear in Matthew Chapter 8 and 13, and 17 through 25. These parables are usually stories of stewards, servants, farmers or householders who demonstrate either good stewardship or bad stewardship by their conduct. These stories always convey the idea that good stewardship is the key to human survival. It is the scientific answer to the scientific question: "What do we need to know and do in order to survive and thrive in this universe?" By these parables, Jesus illustrates what good stewardship is. By healing physical and mental illnesses, Jesus demonstrated good stewardship through his own behavior. His life is an example of good stewardship. Before examining some of the parables that provide clear examples of what good stewardship means, let us look at Chapter 12, at a passage that is extremely powerful in the way it illustrates who Jesus is, what his mission is and how his entire life and death were one consistent demonstration of good stewardship.

Matthew, Chapter 12:

(The text tells the story)

9. And when he had passed on from that place he entered their synagogue. 10. And behold, a man with a withered hand was there. And they asked him, saying "Is it lawful to cure on the Sabbath?" that they might accuse him. 11. But he said to them, "What man is there among you who, if he has a single sheep and it falls into a pit on the Sabbath, will not take hold of it and lift it out? 12. How much better is a man than a sheep! Therefore, it is lawful to do good on the Sabbath." 13. Then he said to the man, "Stretch forth thy hand." And he stretched it forth, and it was restored, as sound as the other. 14. But the Pharisees went out and took counsel against him, how they might do away with him. 15. Then, knowing this, Jesus withdrew from the place; and many followed him and he cured them all, 16. and warned them not to make him known; 17. that what was spoken through Isaiah the prophet might be fulfilled, who said, 18. *Behold, my servant, who I have chosen, my beloved in whom my soul is well pleased: I will put my Spirit upon him, and he will declare judgment to the Gentiles. 19. He will not wrangle, nor cry aloud, neither will anyone hear his voice in the streets. 20. A bruised reed he will not break, and a smoking wick he will not quench, ...* (italics in the text).

What is conveyed to us here in Chapter 12 is truly profound. First, Jesus is once again demonstrating good stewardship by healing a person on the Sabbath. According to Jewish law at that time, it was unlawful to heal on the Sabbath, perhaps because healing was a form of "work." Once again he is illustrating that human needs are more important than religious rituals and institutional customs. We also see in this story a key truth and fundamental reality of Jesus' identity and his mission. The theocratic (political and religious) authorities, the Pharisees, see Jesus as a rabble-rouser and threat to their status and power. Jesus is a genuine threat to the governing class in their eyes, and in the eyes of many of the people, including his own disciples. He is a descendant in the House of David. That means he could legitimately claim to be royalty. He could claim to be the rightful king of Judea and ask people from all classes to support him in a civil war intended to overthrow the Roman occupiers and the Jewish officials that cooperate with Rome. This possibility is recognized by Christian theologians and historians. Given who Jesus was, he could have made a claim to power, and such a claim would have been taken seriously. Many of his disciples expected him to ride into Jerusalem for the Passover celebration on a warhorse, not a donkey, with soldiers and armed civilians supporting his claim to the throne. Some wanted him to make the crowded city the stage for a declaration of war -- the war that would lead to victory for their version of a warrior "Messiah." He rode in on a donkey, because that is what a servant would be riding. He allowed himself to be taken captive in the Garden of Gethsemane and convicted of a religious crime, "blasphemy," so that the religious authorities could discredit him. Pilate, the Roman legal authority, did not see Jesus as having done anything illegal. The charge of "blasphemy" was not taken seriously by the Romans. That's why Pilate "washed his hands" and turned Jesus over to the religious authorities who wanted him to be punished and humiliated. That was simply a political move by Pilate. The Romans wanted the cooperation of the religious authorities, so he released Jesus to them. In this way Jesus was punished by both the Roman authorities and the Jewish authorities, but Pilate himself put on record that he did not see Jesus' behavior as criminal. Some of the Jewish people supported Jesus, who they knew as a just man and benevolent healer, so Pilate avoided being the villain in the people's eyes by stating on the record that he saw no criminal offense. But look again at the prophecy of Isaiah. It says that a servant shall be sent to Israel, and:

"He will not wrangle, nor cry aloud, neither will anyone hear his voice in the streets. 20. A bruised reed he will not break, and a smoking wick he will not quench, ..."

This means that Jesus will not cause more suffering to the people who are already suffering the burdens of the Roman occupation. He will not foment civil unrest nor take any steps to promote a civil war. Even though he has a legitimate claim to the political power and wealth of kingship, he chooses death at the hands of his enemies, and his exquisitely unswerving commitment to good stewardship. He prefers the treasures of wisdom and a life of service to human needs rather than the emptiness of material wealth and political power. Who has put the welfare of others ahead of himself as devotedly as Jesus? Look again at the stories of the end, in all four gospels, when he is taken captive and tried and whipped, publicly humiliated, and executed. He could have escaped this easily. He "withdrew" into the desert many times, including times when he was aware that the authorities conspired against him. Look at the many passages that refer to Jesus being in direct conflict with the authorities.

> Matthew 9: 34: But the Pharisees said, "By the prince of devils he casts out devils."

> Matthew 12: 14-15: But the Pharisees went out and took counsel against him, how they might do away with him. ... 15. Then, knowing this Jesus withdrew from the place;

> Matthew 15: 12-14:
> 12. Then his disciples came up and said to him, "Dost thou know that the Pharisees have taken offense at hearing this saying?" ... [Jesus responded] ... 14. Let them alone; they are

blind guides of blind men."

Matthew 16: 20: Then he strictly charged his disciples to tell no one that he was Jesus the Christ. And Matthew 20: 18: Behold, we are going up to Jerusalem, and the Son of Man will be betrayed to the chief priests and the Scribes; and they will condemn him to death; ...

Matthew 21: 15-17: (The Cleansing of the Temple)
15. But the chief priests and the Scribes seeing the wonderful deeds that he did, and the children crying out in the temple, and saying "Hosanna to the Son of David," were indignant. 16. and said to him, "Dost thou hear what they are saying?" ... 17. And leaving them, he went out of the city to Bethany and stayed there. [Jesus had relatives and friends in Bethany.]

The Scribes and Pharisees asked Jesus "by what authority" he was teaching the people (Matthew 21: 23-27). Jesus was, to them, a resented competitor for the attention and the genuine respect of the people. He spoke by the authority of truth respected by the people, but he did support the institutional authority of the temple, the priests, Scribes and Pharisees. That Jesus was seen as religious and political opposition is clear in Matthew 21: 42-46 (they sought to lay hands on him):

46. and though they sought to lay hands on him, they feared the people, because they [the people] regarded him as a prophet. [However, ...]

Matthew 26: 3-5: Then the chief priests and the elders of the people gathered together in the court of the high priest,

who was called Caiphas, 4. and they took counsel together
how they might seize Jesus by stealth and put him to death.
5. But they said, "Not on the feast, or there might be a riot
among the people."

Jesus had friends, even Roman soldiers. He could have avoided capture.
Even the Jewish court of religious law, called the "Sanhedrin," did not really
want to punish him personally. They had to do so in order to publicly
respond to Jesus' challenge to the legitimacy of their religious authority. It is
very important for Christians to remember that Jesus is the quintessential
rebel, one who questions authorities. He was the master. He did not oppose
the authority of civil law. He said that people should pay their taxes and
practice non-violence. His life is an example of non-violence. The Jewish
authorities saw him as corrupting the youth, exciting the youth to lose
respect for their elders, and those who hold political office. His accusation
and condemnation is practically the same as the means by which the Greek
authorities condemned the philosopher Socrates, whose determined pursuit
of ultimate truth and persistent practice of "asking questions" drove them to
anger and finally to their disturbed desire to kill the man who questioned
their authority. The executions of Socrates and of Jesus are best understood
as two occurrences of the same type of event in slightly different cultures.
One culture was Greek and rather secular, the other was Jewish and
theocratic. But both had the same distaste for common people who ask too
many questions. This reality is extremely important for understanding what
is happening in America today, where authoritarians who present themselves
to the people as being "conservative Christians" argue that to question the
authorities is unpatriotic, even supportive of the nation's external enemies.
Jesus knew that his questioning of the authorities would lead to his arrest,
trial and execution. He knew this as a common pattern of human behavior
throughout history. And so we can see here who Jesus is. His mission is to
describe good stewardship in words and illustrate good stewardship with his

life and voluntary death. He voluntarily accepts mistreatment and even death at the hands of cruel, authoritarian hypocrites rather than be the cause of more violence and suffering for the people. This is an extreme example of good stewardship, given freely, so that the extreme importance of good stewardship would be made clear to us, all of us.

Now let's look at Chapter 13, a wonderland of effective metaphors for "the kingdom of heaven." We have already studied the "Parable of the Sower" in Chapter Three, where the kingdom of heaven is like a farmer who casts seed everywhere and some settles on good ground. The Parable of the Weeds, Matthew 13: 24-30: (a householder)

> 24. Another parable he set before them, saying, "The kingdom of heaven is like a man who sowed good seed in his field; 25. but while men were asleep, his enemy came and sowed weeds among the wheat, and went away. 26. And when the blade sprang up and brought forth fruit, then the weeds appeared as well. 27. And the servants of the householder came and said to him, 'Sir, didst thou not sow good seed in thy fields? How then does it have weeds?' 28. He said to them, 'An enemy has done this.' And the servants said to him, 'Wilt thou have us go and gather them up?' 29. 'No,' he said, 'lest in gathering the weeds you root up the wheat along with them. 30. Let both grow together until the harvest; and at harvest time I will say to the reapers: Gather up the weeds first and bind them in bundles to burn; but gather the wheat into my barn.' "

The Christian clergy usually explain this as a metaphor where the weeds represent immoral people and the wheat represents good people. I have heard many sermons on a Sunday morning using this parable as a means to

28

rouse the congregation to assess their moral selves -- are you wheat or weed? My perception is that the wheat represents good stewards, and the weeds represents those who are unable to be good stewards either because they don't want to be or because they do not understand the importance of the invitation to be a good steward. In any case, we see once again that the kingdom of heaven is a "selector." Like someone shopping for food for the home or someone buying goods for a business enterprise.

The Parable of the Leaven in Dough, Matthew 13: 33:
33. He told them another parable; "The kingdom of heaven is like leaven, which a woman took and buried in three measures of flour, until all of it was leavened."

In the Parable of the Leaven in Dough, the kingdom of heaven is described as being "mixed" or thoroughly kneaded into all of the "dough." To be consistent with my viewpoint regarding panspermia (life is disbursed throughout the universe), and the Parable of the Sower discussed in Chapter Three, the "dough" is actually the universe, or at least a significant section of the universe, where the kingdom of heaven casts seeds to float in all directions, or "kneads leaven" in all directions, throughout the universe. In this case, "leaven" being yeast or like yeast, is a kind of living seed also, that can grow and evolve, make flour into living bread.

The Treasure and the Pearl:
44. "The kingdom of heaven is like a treasure hidden in a field; he who finds it hides it, and in his joy goes and sells all that he has and buys that field. 45. "Again, the kingdom of heaven is like a merchant in search of fine pearls. 46. When he finds a single pearl of great price, he goes and sells all that he has and buys it.

The Fisher's Net:

47. "Again, the kingdom of heaven is like a net cast into the sea that gathered in fish of every kind. 48. When it was filled, they hauled it out, and sitting down on the beach, they gathered the good fish into vessels, but threw away the bad. 49. So will it be at the end of the world. The angels will go out and separate the wicked from among the just, 50. and will cast them into the furnace of fire, where there will be the weeping and the gnashing of teeth."

With these metaphors in Chapter 13, three important concepts are converging:

One: The kingdom of heaven is like a shopper, one who selects what to buy.

Two: The kingdom of heaven values what is best, such as a hidden treasure or a pearl of great price. The planet Earth is a pearl of great price, because it is a water planet that supports life in abundance, millions of species, not just a few.

Three: Those who are bad stewards and are not selected experience painful loss when they realize the importance of the invitation that was extended to them and that they failed to respond properly. They lost the greatest opportunity possible, the opportunity to be good stewards and members of the kingdom of heaven. We are usually told by the clergy that individual human beings will be selected if they are good stewards as individuals. We cannot be certain this is true. It appears to me that we all win or lose together, as a civilization and as a species. The kingdom of heaven may not be selecting individuals, but rather selecting those species of intelligent beings or technological animals that are truly good stewards. That would mean our fate is collective, not individual. I do not claim to know which it is. I suspect that it could be both, that in the process of selecting individuals who are good stewards, the kingdom of heaven is also selecting an entire species, a species that "lives" in the traits of the individual.

The total impact of the parables in Matthew Chapter 13 is to convey to us that the kingdom of heaven spreads primitive forms of life throughout the universe and then observes where life settles on "good ground" or planets where it survives and thrives. Then the kingdom of heaven observes where technological animals and intelligent beings evolve, and watches for the development of any form of "good steward" species, that is, intelligent beings who actively take care of the life and life-supporting environment that is within their reach. These good stewards are what the kingdom of heaven is looking for, because "more laborers" are needed for the "harvest," the promotion and protection of intelligent life in the universe -- and all life in the universe. Note again that the process whereby the kingdom of heaven selects good stewards is the same as when a human shopper buys apples or fish, or a human shopper in business buys tools or parts or hires a crafts person. The rejection of inferior products is natural, coldly logical and neutral morally. It has nothing to do with anyone feeling the need to punish anyone. My perception is that the rejection of "bad stewards" is the same thing as a parent telling her children that certain people should not be allowed to enter the house, or upon the property, because they are bad and dangerous. That is just natural, common sense. That is why I say that the selection of good stewards and the rejection of bad stewards is probably a "natural consequence" and simply the way the universe works. It does not necessarily require the intervention or deliberation of any Supreme Being or God. With or without God in charge, the universe does not select bad stewards by the same common-sense logic that a smart shopper does not buy bad apples. We can be encouraged, though, by the parables in Chapter 13. They tell us that good stewards are deemed to be of great value. They are like a "hidden treasure" and a "pearl of great price." That is our real value, if we are good stewards. We will not be left waiting at the marketplace, not thrown out with the rotting fruit. The kingdom of heaven would "sell whatever it has" in order to have us, the hidden treasure and pearl of good stewardship. If you are still not sure that being a good steward is the meaning of good behavior, and what Jesus is telling us the kingdom of heaven wants from us, I will show you some confirmation shortly, in the next Chapters. Jesus made it very clear that good stewardship is the form of good behavior that counts.

Other parables about the kingdom of heaven and stewardship include:
The Unmerciful Servant (in summary): When a servant cannot pay a debt to his master, he begs for compassion from his master and receives fair treatment, but he in turn is impatient and punitive toward his subordinates. In the end, he gets punished for his cruelty. Jesus reminds us that we will be treated by the kingdom of heaven the same way we treat others.

Parable of the Two Sons (in summary): A father, with a vineyard of course, asks one of his sons to work in the vineyard, and he says "No." The father asks his second son, and the second son says "Yes." But the first son thought that he should honor his father's request and even though he had said "No" he went and worked in the vineyard. The second son, although he had said "Yes" did not go and do any work in the vineyard. Jesus uses this story to emphasize that the first son who said "No" but did his father's will is the good son, the good steward, the one who will pass the judgment test because we are judged by our conduct, not what we say or what we say we believe.

The Marriage Feast (in detail): Matthew 22: 1-14:
1. And Jesus addressed them, and spoke to them again in parables, 2. saying, "The kingdom of heaven is like a king who made a marriage feast for his son. 3. And he sent his servants to call in those invited to the marriage feast, but they would not come. 4. Again he sent out other servants, saying, 'Tell those who are invited, Behold, I have prepared

my dinner; my oxen and fatlings are killed, and everything is ready; come to the marriage feast. 5. But they made light of it, and went off, one to his farm, and another to his business; 6. and the rest laid hold of his servants, treated them shamefully, and killed them.

7. "But when the king heard of it, he was angry; and he sent his armies, destroyed those murderers and burnt their city. 8. Then he said to his servants, "The marriage feast indeed is ready, but those who were invited were not worthy; 9. go therefore to the crossroads, and invite to the marriage feast whomever you shall find. 10. And his servants went out into the roads, and gathered all whom they found, both good and bad; and the marriage feast was filled with guests.

11. "Now the king went in to see the guests, and he saw there a man who had not on a wedding garment. 12. 'Friend, how didst thou come in here without a wedding garment?' But he was speechless. 13. Then the king said to the attendants, 'Bind his hands and feet and cast him forth into the darkness outside, where there will be the weeping, and the gnashing of teeth. 14. For many are called, but few are chosen.'"

Here are the profoundly important concepts conveyed by this parable:

-- The kingdom of heaven extends an invitation to us, and to many, to become good stewards and come to the joyful work, which is like a wedding celebration, of taking care of life in the universe. This is a very important invitation to participate in very important work.

-- Some of those invited clearly did not understand the great importance of the invitation. This invitation is not to be lightly dismissed. Not only did some of the invitees ignore the invitation, some were rude and killed the servants sent out to deliver the invitation message. This may seem an extreme response to an invitation, and thoughtlessly violent behavior, which it is exactly. This represents an accurate statement by Jesus describing how people have treated the prophets throughout ancient history. The prophets are those who advised the people that they were behaving badly and that the people and the government needed to pay more attention to social and economic justice, the moral behavior that counts. Such prophets were usually respected by many but often mistreated, tortured and killed by the political or institutional authorities. This kind of response to a generous invitation warrants a serious consequence, and we see that the consequence described in this parable is that the king killed the violent invitees and burnt their city. For us, who are invited to be good stewards and join the kingdom of heaven, this would be analogous to the kingdom of heaven killing us and burning our cities, or "burning" our entire civilization, if we continue to both refuse the invitation and mistreat those who deliver that invitation to us.

-- Next, this parable tells us that there are others who can be invited in our place if we are so stupid that we do not understand the significance of the invitation to be good stewards.

-- Next, we are shown that those others who can be invited in our place also need to show that they understand the importance of the invitation and their duty to show respect for the kingdom of heaven who has extended the invitation. Those who do not wear a wedding garment to the marriage feast are irresponsible because in the culture where Jesus told the parables, a wealthy family would send wedding garments to those invited to a wedding so that no one who was invited would be embarrassed by not having suitable clothing for the celebration. Anyone invited was expected to wear the garment provided by the host and thereby arrive as a guest among equals and show off the generosity and taste of the host. To receive a wedding garment from the host and then go to the wedding celebration not wearing that wedding garment was an insult to the host, and a sign again that the one invited did not understand the importance of the invitation. Or, it could mean the person attending was not actually invited. So, this parable is not just an exaggeration of the importance of social skills as a kind of moral or good behavior, but a metaphor that is intended to convey the extreme importance of the fact that those who are invited to join the kingdom of heaven need to have an accurate understanding of what that invitation really means. Otherwise, we will only show that we do not understand and we are not really ready to be good stewards and we are not ready to join the kingdom of heaven. If you are not invited, you cannot make yourself a legitimate guest just by showing up. Being uninvited or unprepared gives us very painful results, described to us as having our hands and feet bound (losing our freedom) and being cast outside into the darkness (being outsiders and ignorant and uninformed) and "weeping and gnashing our teeth" when we finally come to realize that we missed the greatest opportunity in the universe, and it is too late to change our indefinite future as outsiders in the dark. This may appear to be a harsh reality, but if this is the reality of how the universe works -- I believe it is -- then it was fair for the kingdom of heaven to send us messengers to let us know. The phrase "many are called" means many species are invited to be good stewards. The phrase "few are chosen" means only a few of those invited make the grade. So, once we grasp what the invitation means, the rest is up to us. This is scientific information about how the real, physical universe works. This is not moral philosophy that applies to an "inner life" of contemplation.

The Need for Vigilance (in summary): Matthew 22: 46-51:
Jesus conveys the importance of the invitation and expectation of the kingdom of heaven by describing the good and faithful servant as one who gives the subordinate servants their food and meets their needs even while the master of the house is gone.

Matthew 22: 46-51:

46. "Blessed is that servant whom his master, when he comes, shall find so doing. 47. Amen I say to you, he will set him over all his goods. 48. But if that wicked servant says to himself, 'My master delays his coming.' 40. and begins to beat his fellow servants, and to eat and drink with drunkards. 50. the master of that servant will come on a day he does not expect, and in an hour he does not know. 51. and will cut him asunder and make him share the lot of the hypocrites. There will be the weeping and the gnashing of teeth."

-- Clearly, being "cut asunder" is being made an outsider in the dark, and then experiencing the loss of the priceless opportunity to be a permanent and respected member of the household, or of the big household in the kingdom of heaven.

Parable of the Ten Virgins (in summary):
In this parable, Jesus presents the story of ten young women invited to a wedding feast and how they need to understand the importance of the invitation. The invitees demonstrate their understanding by their behavior. As evening falls, the young women are expected to provide light for themselves and the other guests with lamps filled with oil. Five foolish virgins came with lamps but no oil. Five wise virgins came with lamps and oil. Those who had the good sense to bring a supply of oil were not obligated to share with those who had not. Those without oil had to go out looking to buy it. Later, when they returned to the wedding celebration, the bridegroom said to them "Amen I say to you, I do not know you." And they were locked outside. Again, those who were not prepared to meet their responsibilities, not prepared to be good stewards, were locked out permanently. The idea that those invited must understand the meaning of the invitation is driven into our heads like a nail being pounded by the repeated blows of a hammer

into a block of wood. Your head, and mine, are blocks of wood. Is the message sinking in? Jesus is a teacher of primitive students. He uses repetition as a teaching tool.

Parable of the Talents (in detail): Matthew 25: 14-30.

In Matthew 25 we have the most important document in the world. Here we will take a good look at how it describes good stewardship and the concise way in which Jesus conveys to us that what we need to do is *nurture the vulnerable and weak*, so that they can become strong. And, of course, the strong need to be free, or liberated, in order to do the good work that they can do for the benefit of the community, of society, of the nation, and of the world. Later in Chapter Six, we will look at Matthew 25 again to explore another aspect of Jesus' teaching, the reality that we will be judged according to our conduct. Matthew 25 tells us that it is our behavior that reveals what we believe, not our words or prayers or rituals or membership in any organization.

The second most important document in the world is Matthew 23, which we will look at in Chapter Seven, which is about three unforgivable sins. Matthew 23 is the most important document in the world in the field of political and social science, because it describes how people behave when they have an unequal social status or political power. The cynical, hypocritical and authoritarian behavior described in Matthew 23 is not *always* the way people with power behave, but every human being is *at risk to behave* in this way when exposed to power. This is why the founding fathers of the United States of America, who embody a moment of wisdom in human history, liked to say "Power corrupts and absolute power corrupts absolutely."

Parable of the Talents (14-30):

14. "For it is like a man going abroad, who called his servants

and handed over his goods to them. [This "it" must be our friend, the kingdom of heaven.] 15. And to one he gave five talents, to another two, and to another one, to each according to his particular ability, and then he went on his journey. 16. And he who had received the five talents went and traded with them, and gained five more. 17. in like manner, he who had received the two gained two more. 18. But he who had received the one went away and dug in the earth and hid his master's money.

19. "Then after a long time the master of those servants came and settled accounts with them. 20. And he who had received the five talents came and brought five other talents, saying, 'Master, thou didst hand over to me five talents; behold, I have gained five others in addition.' 21. His master said to him, 'Well done, good and faithful servant; because thou hast been faithful over a few things, I will set thee over many; enter into the joy of thy master.'

22. "And he who had received the two talents came and said, 'Master, thou didst hand over to me two talents; behold, I have gained two more.' 23. His master said to him, 'Well done, good and faithful servant; because thou hast been faithful over a few things, I will set thee over many; enter into the joy of thy master.'

24. But he who had received the one talent came and said, 'Master, I know that thou art a stern man; thou reapest where thou hast not sowed and gatherest where thou hast not winnowed; 25. and as I was afraid, I went away and hid thy talent in the earth; behold thou hast what is thine.' 26. But his master answered and said to him, 'Wicked and slothful servant! thou didst know that I reap where I do not sow, and gather where I have not winnowed? 27. Thou

shouldst therefore have entrusted my money to the bankers, and on my return I should have got back my own with interest. 28. Take away therefore the talent from him, and give it to him who has the ten talents. 29. For to everyone who has shall be given, and he shall have abundance; but from him who does not have, even that which he seems to have shall be taken away. 39. But as for the unprofitable servant, cast him forth into the darkness outside, where there will be the weeping, and the gnashing of teeth.'

Jesus is repetitive on the issue of stewardship. He is constantly talking to us about good servants and bad servants. He tells us again and again that those who do not possess the qualities of a good steward will be "cast outside" where there is "weeping and gnashing of teeth." Concentrate first on the key messages here, and the proper interpretation of the metaphor of money and earning a profit. The money and profit represents the value of productivity, the worth of constructive labor, the value added to what one already has by virtue of proper care alone or further by improvement or enhancement. It is really another way to declare the principle of stewardship, a way that is more easily understood by a banker or investor who takes the risk of shipping goods across the sea or across the desert to be traded, as opposed to the previous stories of farmers and householders. Such a trader or businessman could be described by the common people as one who "reaps where he does not sow" or "gathers where he does not winnow," because one who makes his living buying and selling goods does not add value to goods with his own hands, as the common farmer or craftsperson does. However, in fact the trader or storekeeper makes his living by applying his own judgment as to the right price for an object, by assessing the demand and the supply of goods, and what is worth holding until the price is right, and what is best sold as soon as possible at any reasonable price offered, and by advocating for the value of the product.

It is an error to interpret this parable as a strong defense of capitalism, or of banking and charging interest. Although there are benefits for society in systems of banking and the management of money, and of course in the management of productive tools and property, this parable is not about capitalism alone. The parable of the talents is about something more basic and generic. The talents represent the things or property or even the personal skills within reach of each of the servants. One might say that each servant has an "allotment" or a given quantity of goods, or skills, to work with. It really is as simple as saying: "What you do with what you've got is what really counts." This is folk wisdom, not a deep or complex dissertation on banking or industrial production. This parable says that each of us must take the risk involved in our own efforts to be actively productive, to be constructive rather than passive or destructive, to add something to what we have. It does not mean only to add value to the property we have, but to add value to ourselves, to educate ourselves for example, to act in order to learn, to learn by doing, and then to do new things because we have learned that we can do things that are new to us. The parable of the talents is not a reference to high finance or stock brokers or the accumulation of industrial capital. It is a reference to the importance of doing something productive with whatever one's resources are. And, simple enough, Jesus is giving us another story to help us grasp the meaning of stewardship. Doing something productive with what we receive, however great or small it may be, is good stewardship.

What is really important here is to avoid the mistake that so many make. Even the traditional theologians focus on the fate of the "bad steward." He must be punished for being lazy, or a coward, or for being afraid of the master, and for being afraid to take the risk of bringing the one talent into the marketplace and perhaps losing it. This servant does not appear to me to be "bad" in the usual sense of the term. He has done nothing wrong. The problem is that he has done nothing good either. This is the essence of Jesus' teaching, the key difference between the Ten Commandments of the Old Testament and the Primacy of Stewardship of the New Testament, that we do not meet the requirements of the kingdom of heaven simply by observing the Ten Commandments. The Ten Commandments are a list of prohibitions, bad things to not do. One who "has eyes to see" can see that Jesus is teaching us that in order to be worthy of the kingdom of heaven we must not only avoid doing what is bad; we must also do what is good. And stewardship, productivity, adding to the value of what one has, is doing good. The best lesson of the Parable of the Talents comes not from focusing on the losses of the bad steward, but from focusing on what happens to the two other stewards. One has five talents, and the other has only two. But they each have the same fate:

> "Well done, good and faithful steward. Because you have been faithful over a few things, I will set thee over many. Enter into the joy of thy master."

These phrases are laden with meaning, with the meaning of what it means to join the kingdom of heaven. First, the servant with five talents and the servant with two talents become equals because they are both good stewards. Because the servant with five talents had more to begin with, he is expected to produce more. His reward is not greater than the reward of the servant with two talents. Both are advised of the reward for good work: more responsibility. Both are advised that they will be set over more goods because they have demonstrated that they can be trusted. Both are invited to "enter into the joy of your master," which means that whatever happiness you think a "master" finds, is now yours. You are one of us, one of those who takes the risks involved in active productivity and succeeds.

The steward who loses everything is really only the same as the virgins who did not bring oil with their lamps, the same as those who were invited to the marriage feast but did not go, or who assaulted the servants delivering the invitations; the same as the vine dressers who did not take proper care of the vineyard; the same as the servant who neglects the needs of his subordinates and "eats and drinks with drunkards." The steward who buried the one talent entrusted to him is not evil and not even destructive. What he lacks is understanding. His deficiency is that he does not fully understand what it meant that valuable property was entrusted to him. This is the same as us not knowing the value of the planet Earth that has been entrusted to us. He does not know the value of what he has. He did not understand his responsibility to take good care of it, and he did not understand that to really take good care of it means to exercise his own ability to add value to it or find the opportunity to have someone else add value to it.

The true meaning here, beneath the metaphorical setting of money and interest and profit, is that the "bad" steward did not understand the invitation or the opportunity that was given to him. He did not know that he could have added value to what he had and by so doing he would be a "good steward" worthy of greater responsibilities and worthy of finding the same happiness, whatever that may be, of his master, who is, be assured, the kingdom of heaven. The theme of the importance of good stewardship is repeated many times, probably because repetition is the most basic of teaching methods.

Thinking back about the good farmers and householders who were productive in Jesus' parables, we can see that the true purpose of The Parable of the Talents is to illustrate that a banker or a trader who does not take any risks to make a profit is like a farmer who has land but does not plant any seeds, or a man with a vineyard who does not hire laborers to take care of it and then gather the harvest. To conclude that The Parable of the Talents defends the morality of capitalism implies that the parables about the weeds and the sower and a vineyard are all intended to defend the morality of farming. That does not make sense. The occupations of farming, banking, trading, fishing and so forth are in themselves morally neutral. Morality applies to the person, not the occupation. A banker or a farmer, a teacher or a prostitute can apply moral principles to the manner in which they meet the demands of human society. A banker could be asked to hide a theft. A farmer could be asked to grow illegal crops. A teacher can be told to teach lies. A prostitute can be told to drug a man so that he can be beaten and robbed. We all work under duress. We all face the challenge of stewardship. Even a thief who does not steal a child's medicine has made an ethical decision.

What property or "treasure" has been entrusted to us? What are we doing with it? By now we should know the answers to these questions. The treasure entrusted to us is the "the pearl of great price," the planet Earth. Again, I am being only as repetitive as Jesus, not more. His mission is to tell us what we need to know. His mission is to alert us to the importance of the invitation that has been extended to us. Are we going to the marriage feast dressed in the wedding garment given to us by the kingdom of heaven? Have we brought oil for our lamps? We have been given a life. Are we adding value to it?

The Last Judgment Matthew 25: 31-40.

31. "But when the Son of Man shall come in his majesty, and all the angels with him, then he will sit on the throne of his glory; 32. and before him will be gathered all the nations, and he will separate them one from another, as the shepherd separates the sheep from the goats; 33. and he will set the sheep on his right hand, but the goats on the left.

34. "Then the king will say to those on his right hand, 'Come, blessed of my Father, take possession of the kingdom prepared for you from the foundation of the world; 35. for I

was hungry and you gave me to eat; I was thirsty and you gave me to drink; I was a stranger and you took me in; 36. naked and you covered me; sick and you visited me; I was in prison and you came to me.' 37. Then the just will answer him, saying, 'Lord, when did we see thee hungry, and feed thee; or thirsty, and give thee drink? 38. And when did we see thee a stranger, and take thee in; or naked, and clothe thee? 30. Or when did we see thee sick, or in prison, and come to thee?' 40. And answering the king will say to them. 'Amen I say to you, as long as you did it for one of these, the least of my brethren, you did it for me.'"

The message here in Matthew 25 is not hidden. Jesus clearly states that when we nurture any human being who is vulnerable and weak due to the circumstances of life that can occur to any one of us, we are nurturing God. It is a sign of one's personal issues if they feel compelled to focus on the fact that some are chosen and others are rejected, or "punished." Jesus does not emphasize punishment. He emphasizes what the scriptures identify as "mercy" and "forgiveness." If we look more closely at the teaching, and focus on the core message of the Gospels, one can see that the references to "mercy" and "forgiveness" are really references to "understanding." The core message in this passage is that Jesus is like us, and we are like Jesus. Amazing as it seems, Jesus is telling us that he is one of us, an "ordinary" human being, and we are one of him. This is what Jesus really says. There are many different forms of examples in the Gospels that state in some way that we are not as different from Jesus as we might feel. We look at him, and we see him healing the blind by touching their eyes, raising people from the dead, or more likely from a coma, and walking on water, healing mental illnesses, and re-attaching an ear that was just cut off of a man's head. How can he do these things and then tell us we are just like him, that he is one of us? Notice, that Jesus does not refer to himself as the "Son of God" but as the

"Son of Man." Since Jesus refers to himself as the "Son of Man," or a son of humankind, he is telling us that he is a human being. We and Jesus are the same species. How is it then, that he has powers and knowledge that we do not? There are many answers to this question, and you will have to find your own. We all need to respect one another's religious beliefs in order to practice the justice and love that is taught to us by Jesus.

I do want to point out that human beings, and other animals, possess an astounding quality that could be the *scientific explanation* as to how and why we are the same as Jesus. Biologists who focus their studies on human genetics have discovered that genes are like power switches or "circuit breakers" in an electrical system. Any home or factory that has electric power will have a control box where there are many circuit breakers so that if there are any problems in one room or one section of the system, the power will be turned off automatically, or it can be turned off manually for that one room or section only. Power is still "on" and available to the rest of the system. Similarly, we have genes that are "turned on" and genes that are "turned off." The genes that are turned on take effect in our bodies, of course, and produce some organic chemical that we need or that gives us some capacity that we use and are aware that we can use. Genes that are "turned off" apparently possess the power to produce some trait in us, but whatever that trait may be, we will not have it physically and we will not be aware that the potential is within us. There are traits that we want turned off, such as the ability to grow a tail, or to grow gills to enable us to breath under water. It may sound like it would be good to have the power to breath under water with gills, but if we had gills we would certainly have a set of problems that would arise from having openings into our throat when we are walking around out of water instead of in it. This possibility may seem to be totally fantastic to you, but it is a fact that a human fetus appears to have gill-like structures temporarily during the fetal development process. We also know that a certain small percentage of human babies are born as hermaphrodites, which means they possess, physically, sexual organs of a female and of a male. This is a form of proof that we all possess some genes that are "turned off" and some that are "turned on." We are designed to have certain female genes turned on and certain male genes turned off, or the opposite, but we are not supposed to have both sets of gender genes turned on with regard to the physical structures of male and female gender. Unfortunately, a reality of genetic development, for all animals, is that development is sometimes, though rarely, misguided due to interference in the normal developmental process. This is why we are so fearful of radiation and organic chemicals that function as "hormone disruptors." Radiation and hormone disruptors can have painfully destructive effects on fetal development. Since it is true that we have genes in our bodies that are "turned off," it is possible that some of those genes would give us healing abilities that Jesus possessed. We may have the potential within us to be able to heal by touching another person, communicate telepathically and thereby understand the thoughts and feelings of another person without hearing them speak orally.

Jesus teaches that the Holy Spirit is within each of us, the same Holy Spirit that is within him. He tells us that it is an unforgivable sin to "blaspheme" against the Holy Spirit. We will take that up in Chapter Seven on "unforgivable sins." Why would we not have these same abilities? Why would such beneficial genes be "turned off?" There is probably more than one answer to the question as to why we have genes turned off, and therefore beneficial powers turned off. I will offer you one possible answer. There is no such thing as a power that is only good, or only bad. We can slice bread with a knife, or slice a throat. Our power to destroy always matches our power to build. The lesson here, a lesson that is certainly visible to me in the teaching and life of Christ, is that technological ability is not an indicator of moral or spiritual progress. A civilization that launches nuclear missiles has not demonstrated any greater moral or spiritual developmental than a civilization that throws stones. Every technological capacity is morally and spiritually neutral. Every technological capacity can be used for either good or evil, and not necessarily in an extreme manner. Our use of technology is often simply to maintain our civilization, not to save it or destroy it in one dramatic action. As our understanding of disease and medicine improves, we are not empowered only to cure and heal, but we are also empowered to kill with biological and bio-chemical warfare. There is no technology that can do only good. Whatever powers a technological animal possesses, the spiritual reality of that being is not expressed in their technology but in how they use their technology. An advanced aircraft can drop bombs or food or medicine or safety equipment. Again, "It is what you do with what you've got that really counts."

How then, is this an explanation of why we have genes that are "turned off?" Since Jesus possessed the ability to heal and cure by touch, or raise a person out of a coma, or heal mental illness, he also possessed the capacity to use these same powers for harm. This sounds strange and discordant at first, because we cannot imagine Jesus deliberately doing harm to anyone. But how could he not possess the ability to do so? If he could heal a person, he could make them ill. If he could raise someone out of a coma, he could put a person into a coma. If he could heal mental illness, he could inflict mental illness. He could make a person blind or deaf, or dumb. Scientifically, if we believe he had the powers to do the constructive things, he had to also possess the powers to do destructive things. Because he was a good shepherd, a good steward, he did no harm. He abided religiously by the Hippocratic Oath. The first directive of the Hippocratic Oath is "Do no harm." This is of course consistent with Jesus' refusal to proclaim himself king and be the cause of a civil war. Do no harm. What if we possessed such powers? Would we be such good stewards that we would "Do no harm?" This is certainly not likely, considering our history, including our immediate history. Our human civilization does not appear to be able to resist the

47

temptation to use its technological powers for purposes of destruction and killing. Therefore, the genes that would give us even greater powers are turned off. If we really become good stewards in the future, then the genes that can give us the same powers as Jesus, the power to heal or make ill, to raise up or strike down, to cure mental illness or induce mental illness, hidden in genes that we already possess, will be turned on. The scientific explanation that I have offered here is an exercise in the "reconciliation of science and religion." It suggests that we may, as Jesus said, become "perfect like our Father in heaven (Matthew 5: 43-48)," in the future. Living organisms have genes that are turned on and other genes that are turned off. Living animals, including humans, have genetic "potential." This could be how the universe works. This could be physical science, not moral philosophy about the individual will.

Chapter Five: The Kingdom of Heaven is Like a Farmer

Jesus repeatedly compared the kingdom of heaven to a farmer or a householder. Christian clergy and theologians of various denominations repeatedly attribute this use of metaphor to the fact that Jesus was speaking to an audience of farmers, fishermen, and ordinary people who knew the routine of managing a household. To me, this is totally mistaken. It is an insult to Jesus and obscures the truth that he was teaching rather than shining a light on it. This interpretation that Jesus chose metaphors for a specific audience makes the appropriateness of the metaphor or the correctness of the comparison, the real teaching tool, less important than the comfort of the audience or the rhetorical skills of the speaker. This speaker is supposed to be, by the way, according to most Christian scholars, God. Whether we believe this is God speaking to us or a very wise man, either way, the speaker is going to be wise enough to choose a comparison or metaphor that is valid. Therefore, my position is that when Jesus said "Heaven is like a farmer," he said it because in fact *the behavior of the kingdom of heaven is like the behavior of a farmer*, and not because he was talking to farmers. And when Jesus said "The kingdom of heaven is like a householder," he said it because in fact *the behavior of the kingdom of heaven is like the behavior of a householder*, and not because he was talking to householders.

Consider this comparison, or metaphor: Penguins swimming in the ocean look like swallows soaring through the air. This comparison will be familiar to many readers. To those who have never seen penguins swimming it may appear to be poetic, an image that is more colorful than it is real. In fact, penguins swimming through water do look almost exactly the same as swallows soaring through the air. The penguins' wings are small, and they often do not appear to be "flapping" very much, but then again, neither do swallows appear to be flapping their wings as they soar rapidly through the air. This metaphor is far more accurate and appropriate than it is colorful or poetic. I certainly did not choose this metaphor because I believe that all or even most of my readers are sailors who have been to the Antarctic seas. If my purpose is to convey some truth, a concept that is accurate and reasonably precise, then this comparison of penguins to swallows is appropriate and effective. It has nothing to do with who is the audience. If you have never seen a penguin swimming, or a swallow flying, you can keep this comparison in mind until you do, and at that time when you have seen both, you will know that the comparison is effective for everyone in the world because it is a meaningful comparison.

I do believe that the kingdom of heaven is truly like a householder and like a farmer, because Jesus said so, and I believe the comparison is meaningful. I believe this interpretation of the farmer and householder metaphors is correct and valid for everyone, valid if you are a lawyer or a soldier or a prostitute or a physician or a gambler or a truck driver. No matter what your occupation, or what routines of the marketplace are familiar to you, it is still true that *the kingdom of heaven is like a farmer*.

Let's look at a couple of parables.

> Matthew 20: 1-16: The Laborers in the Vineyard (quotation of Jesus).
> "For the kingdom of heaven is like a householder who went out early in the morning to hire laborers for his vineyard. 2. And having agreed with the laborers for a denarius a day, he sent them into his vineyard. 3. And about the third hour, he went out and saw others standing in the market place idle; 4. and he said to them, 'Go you also into the vineyard, and I will give you whatever is just.' 5. So they went. And again he went out about the sixth, and about the ninth hour, and did as before. 6. But about the eleventh hour he went out and found

others standing about, and he said to them, 'Why do you stand here all day idle?' 7. They said to him, 'Because no man has hired us.' He said to them, 'Go you also into the vineyard.' 8. But when evening had come, the owner of the vineyard said to his steward, 'Call the laborers, and pay them their wages, beginning from the last even to the first.' 9. Now when they of the eleventh hour came, they received each a denarius. 10. And when the first in their turn came, they thought that they would receive more; but they also received each his denarius. 11. And on receiving it, they began to murmur against the householder, 12. saying, 'These last have worked a single hour, and thou has put them on a level with us, who have borne the burden of the day's heat.' 13. But answering one of them, he said, 'Friend, I do thee no injustice; didst thou not agree with me for a denarius? 14. Take what is thine and go; I choose to give to this last even as to thee. 15. Have I not a right to do what I choose? Or, art thou envious because I am generous?' 16. Even so the last shall be first, and the first last; for many are called, but few are chosen."

This parable conveys one of the most profoundly challenging and profoundly important truths about the kingdom of heaven. Yet again, the Christian clergy conceal its true meaning to avoid conflict with the rich and powerful. The clergy usually interpret this parable as meaning that new members of the church are valued just as much as old members. This is a nice thought, but rarely true. The old members of Christian churches, meaning those who have been active members for a long time, are notoriously possessive of control over the church and all of its policies, including the tiny details such as the types of flowers on display and where and when they are displayed. The older members control what type of bread is used in the communion service and how it is used, what color are the curtains, the placement of informational signs. They own the church, and they are not quick by any means to share that ownership with newcomers. The clergy also use this parable for moralizing and saying that God loves all of us equally, and other such childish avoidance of the disturbing and threatening truth conveyed by this parable. What Jesus is really telling us here is that there is no social or political hierarchy in the kingdom of heaven such as we humans adhere to obsessively on Earth. The kingdom of heaven practices an extreme form of equality, and therefore the kingdom of heaven practices what would be deemed by many humans to be an extreme form of democracy. There is most likely some differentiation, simply to accommodate a division of labor. There is no one in the kingdom of heaven who is "getting paid" more than anyone else. The principle put into practice is "From each according to his abilities, to each according to his needs." If a dollar a day is what is needed to sustain the body and soul, then that is what everyone gets. Not more. Not less. This form of social organization is described on Earth as "communism," or communalism or socialism. In any case, it is a high ideal that we humans only dream of. Jesus has told us, in clear terms, that this serious equality is the rule in the kingdom of heaven, a rule accepted by all of its members. This is why I like to say that "serious democracy requires serious equality."

I believe it is very important to avoid giving this "communist democracy" parable a strictly economic interpretation. It is about both social status and economics, even though this parable is one of those that teaches us what might be called the "economics of Jesus." In sociological terms, this parable tells us that in a just society, there is no divisive economic hierarchy where some are paid hundreds or thousands of times more than others for their "greater" contribution to the work of the society. Those who have more to give in terms of their talents and abilities are expected to give more gladly, and to see what could be called their "greater contribution" as a privilege and honor, not an entitlement to an unequally higher social status. Our Jesus, so enthusiastically described as warm-hearted and forgiving by traditional Christians, is telling us here that the kingdom of heaven is going to make us awesomely equal to everyone else, so Jesus wanted us to know that and be prepared for it. Murmuring against God for not granting you expensive clothing, a limousine, vacation home and restaurant meals is a bad idea.

Review the parable. The laborer who started work early in the day, and felt that he had contributed more, complained that the householder had "put them on a level with us." That resentful laborer complained because the householder had made "unequals" equal in terms of their contributions to the work of the community. Such a policy looks like an advanced form of democracy to me. The kingdom of heaven is busy with its work, which is taking care of the great varieties and great beauty of life in the universe. There is no time for the silliness of social and economic hierarchy. Everyone has the room that they need, the food, the clothing, the shelter, the companionship, the joy of sharing in the tasks, the competence and effectiveness of the community of good stewards. They don't want the illusions and discord that economic hierarchy, or materialism, can cause. All participate in the decision-making and implementation of the decisions. Serious democracy supported by serious equality.

So we see that the kingdom of heaven is looking for laborers to do the work of good stewardship. Are such laborers scarce? Does it take some species of technological animals, such as we humans on Earth, a little longer to catch on to the seriousness of the job offer? Is that us standing around in the market place not sure we are ready to join in the work? At the third hour? At the sixth hour? At the ninth hour? At the eleventh hour? Let's take a look at a second parable of the vineyard. I think it is the same vineyard. And this time, the lesson is even colder: Do the work and do it properly or you will lose everything.

Matthew 21: 33-43:

The Parable of the Vine-dressers (a quotation of Jesus)

33. "Hear another parable. There was a man, a householder, who planted a vineyard, and put a hedge about it, and dug a wine vat in it, and built a tower; then he let it out to vine-dressers, and went abroad. 34. But when the fruit season drew near, he sent his servants to the vine-dressers to receive his fruits. 35. And the vine-dressers seized his servants, and beat one, killed another, and stoned another. 36. Again he sent another party of servants more numerous than the first; and they did the same to these. 37. Finally he sent his son to them, saying, 'They will respect my son.'

38. "But the vine-dressers, on seeing the son, said among themselves, 'This is the heir; come, let us kill him, and we shall have his inheritance.' 39. So they seized him, cast him out of the vineyard, and killed him. 40. When, therefore, the owner of the vineyard comes, what will he do to these vine-dressers?" They said to him [to Jesus]. "He will utterly destroy those evil men, and will let out the vineyard to other vine-dressers, who will render to him the fruits in their seasons."

42. Jesus said to them, "Did you never read in the Scriptures, 'The stone which the builders rejected has become the corner stone; by the Lord this has been done and it is wonderful in our eyes.' 43. Therefore I say to you, that the kingdom of God will be taken away from you and will be given to a people yielding its fruits..."

Here we have what teachers of literature would call a little exaggeration. It is not necessary that the vine-dressers be so violent and evil. They just have to be incompetent. Then again, maybe Jesus is telling us that even some incompetence can be forgiven. It is just that the total incompetence of irrational violence and complete disrespect for the mission of the owner of the vineyard, who wants to enjoy the abundance of the fruits produced by the vineyard, is not forgivable. The extreme betrayal committed by these vine-dressers does suggest, to the alert reader, that if the vine-dressers had been respectful, and diligent, but failed to do a good job due to some lack of knowledge or skills or experience, they might still have a job. Loyalty to the task and just behavior are perhaps the most important qualifications for any hired help.

Still, the message of this parable is otherwise cold and sharp: betrayal of the duty to take care of life and be respectful of all persons is unforgivable failure. The violence and selfishness of the bad vine-dressers is what we would usually call a "disqualification." I would apply this same standard to the overall behavior of the human species during modern times. We have done wonderful things. We have built dams and invented medicines and vaccinations and systems of electrification, systems of roads, improved transportation, produced great works of art and science. Jesus is telling us in this parable that if we kill innocent people, have civil wars and force children to kill and destroy, use bombs to tear off the limbs of children and adults who are trying to live in peace, that, my dangerous friends, is a "disqualification." We have some of the skills required of a "good steward."

Sometimes we behave as though we are good stewards, but the rules and regulations of the universe disqualify a species when they engage in irrational violence. This is similar to how we disqualify a champion athlete who has the highest level of ability, and defeats all opponents, but did so while taking illegal drugs. The athlete's "addiction" to a drug is like the human addiction to violence. We still have what it takes, and we are not necessarily eliminated permanently from being candidates for "good stewardship" and membership in the kingdom of heaven. It's just that for the present we are not eligible.

Our achievements are not counted in the official record, and we are not admitted into the fellowship of good stewards.

To give another example. Let's say you have a wonderful watchdog. He is warm and gentle and loving to every member of your family. He is smart and certainly competent as a watchdog, alerting you to every suspicious sound and to odors and movement only he can detect. But, one evening, a friend arrives unexpectedly and her young child runs toward the house joyfully. Your watchdog, let out for a few minutes, attacks the child and destroys her right arm.

What happens next to your dog? There are some people who would forgive the dog. They are not the kingdom of heaven. In the parable of the vine-dressers, Jesus has informed us of how the universe works. The irrational violence of war is unacceptable even when it is committed by a dumb animal. This was Jesus' mission, to inform us how the universe works. We are now informed. If we find ourselves cast outside and we are weeping and gnashing our teeth because of the opportunity we lost, as is described further by Jesus in Matthew Chapters 24 and 25, we cannot say, "We didn't know." What does it mean to be a "laborer for the harvest?" It doesn't mean that we are going to be harvested like potatoes and included in a popular recipe. It means that a laborer is a good steward, an individual or a civilization that is a good steward, one who participates in taking care of life in the universe so that there is an abundance of beautiful, useful, thriving life.

The kingdom of heaven is like a farmer, and farmer's like to see their fields in full bloom. Think of "amber waves of grain... fruited plains." They like to see a rich and high quality harvest. This is what healthy, intelligent beings like. They like abundant life. They like to heal, nurture, build, restore, maintain, and liberate life. There is a lot of work to do. There are positions to be filled.

Matthew 9: 35-38: 35. And Jesus was going about all the towns and villages, teaching in their synagogues, and preaching the gospel of the kingdom, and curing every kind of disease and infirmity. 36. But seeing the crowds, he was moved with compassion for them, because they were bewildered and dejected, like sheep without a shepherd. 37. Then he said to his disciples, "The harvest indeed is great, but the laborers are few. 38. Pray therefore the Lord of the harvest to send forth laborers into his harvest."

Are these the exact words of Jesus? We cannot be certain. Look at the general idea being expressed. There is life in the universe that needs care. Workers, good stewards, are needed to provide that care. This is what we, technological animals, could become. We have the ability. Jesus' mission and teaching is the best evidence we have that a superior civilization has taken an interest in us, just as we take an interest in the many forms of life that we find on our planet. Their interest in us should be neither puzzling nor frightening. They don't want to eat us. They want us to grow up and help them take care of life in the universe. Are we prepared for this job opportunity?

Chapter Six: Judgment According to Conduct

Jesus tells us with extreme clarity that we will each be judged (or our species will be judged) according to our conduct. This means, of course, we will be evaluated on the basis of what we do, meaning not only what we do occasionally but by our *patterns of behavior*. What do we do regularly? That is the question. What purposes, what knowledge and beliefs are revealed by our actions, by our ongoing behaviors? That is the question.

This may seem trivial and obvious at first. However, think about how our civilization operates. Hundreds of millions of people talk and act as though they really believe that they are going to be evaluated, or judged, on the basis of what they say they believe. The fundamentalists of all religions, in particular the Christian Fundamentalists and the Moslem Fundamentalists, two of a kind, yell and scream their version of what they think God wants to hear, such as "I have a personal relationship with Jesus," or "God is Great! There is no God but God!" The God of these Abrahamic religions is supposedly a very wise old man. If this Christian/Islamic God is a very wise old man, he knows that what people say they believe does not establish what they really believe. What we proclaim we believe, and the religious rituals we observe and the religious institutional authorities we uphold on the Earth, do not reveal to God what we really believe. What we do shows God what we really believe. Our patterns of behavior establish what we really believe. God is like a child who does not learn from hearing what we say, but who learns who we are from what we do, in particular how we treat others. This is exactly what Jesus said to us in Matthew 25 quoted in Chapter Four. We shall be judged in accordance with our behavior toward others. However you render justice to others is how the kingdom of heaven will render justice to you. How and why hundreds of millions of people still think they can be saved by Jesus or protected by Allah just by saying the magic words may be complex. I am sure that every individual's deeply felt need to be included in human society, and the reciprocal demand for conformity that society imposes on everyone, is the greatest cause. We are communal, social beings. We hate being isolated, rejected, alone. In order to belong, we will say almost anything, even watch an innocent person be tortured and executed so that we can continue to live among our neighbors while doing our best to hide the fear that we might be next. I have already discussed the Parable of the Two Sons, where the one who says "No" but does the work requested is the one who has done the will of his father, and therefore the one who is the good steward. The son who said "Yes" but did not act in accordance, did not do the work requested, is the bad steward who is judged unworthy. This discussion could be pursued at great length, but I believe it is up to the reader to examine this important concept on their own.

Let me just say that I see this issue as what I like to call "salvation by membership." Christians and Moslems and many other religions have "teachers" or "mullahs" or "evangelists" who tell people that their souls will be saved if they "join" the religious organization. Being a member will get you the acceptance, approval and protection of God. Don't fall for that nonsense. We cannot be saved or admitted into the kingdom of heaven based on membership in any human organization or institution. Each of us, even our entire species, will be judged according to our conduct. That is a law governing how the universe works. Rather than just repeating this important fact, let's look at the ways in which Jesus conveyed this vitally important message to us.

I find Jesus, our wonderful and repetitive teacher, telling us that we will be evaluated according to our conduct, and not our professions of faith, at least twelve times. Here are a few.

Matthew 5: 43-48:

43. "You have heard that it was said, 'Thou shalt love they neighbor, and shalt hate thy enemy.' 44. But I say to you, love your enemies, do good to those who hate you, and pray for those who persecute and calumniate you. 45. so that you may be children of your Father in heaven, who makes his sun to rise on the good and the evil, and sends rain on the just and the unjust. 46. For if you love those that love you, what reward shall you have? Do not even the publicans do that? 47. And if you salute your brethren only, what are you doing more than others? Do not even the Gentiles do that? 48. You therefore are to be perfect, even as your heavenly Father is perfect."

Matthew 7: 21:

21. "Not everyone who says to me, 'Lord, Lord,' shall enter the kingdom of heaven, but he who does the will of my Father in heaven shall enter the kingdom of heaven."

Matthew 12: 46-50:

46. While he was still speaking to the crowds, his mother and his brethren were standing outside, seeking to speak to him. 47. And someone said to him, '"Behold, thy mother and thy brethren are standing outside, seeking thee." 48. But he answered and said to him who told him, "Who is my mother and who are my brethren?" 49. And stretching forth his hand towards his disciples, he said, "Behold my mother and my brethren! 50. For whoever does the will of my Father in heaven, he is my brother and sister and mother."

Matthew 16: 26-27:

26. For what does it profit a man, if he gain the whole
world, but suffer the loss of his own soul? Or what will a
man give in exchange for his soul? 27. For the Son of Man
is to come with his angels in the glory of his Father, and
then he will render to everyone according to his conduct."

That last passage seems rather direct. Next, in Matthew Chapter 17 we have
the Parable of the Unmerciful Servant. In Chapter 21 we have the Parable of
the Vine Dressers and in Chapter 22 we have the Parable of the Marriage
Feast. In Chapter 23, we have the incredibly important dissertation on the
hypocrisy of those vested with religious and political authority. In Chapter
24 we have the Parable of the Ten Virgins and the reference to the behavior
of a bad servant. In Chapter 25 we have the Parable of the Talents and the
description of "The Last Judgment." Again and again Jesus tells us that we
shall be evaluated according to our conduct. Still, anyone who observes
human behavior can see that the concept of salvation by words, or by
membership, prevails among many and even among Christian and Muslim
teachers. This is so, I think, because it is so much easier to cry out "I love
God," or sing a hymn, or bow in prayer or participate in a customary ritual --
in view of one's friends and neighbors -- than it is to do what Jesus did: sit
down as an equal with the poor and the "sinners," show understanding and
mercy for the tortured heart of the criminal, forgive those who are different
from you and ridicule your beliefs or your customs, nurture the weak and
vulnerable instead of exploiting them for labor or to assign oneself a more
"noble" social status.

It is so much easier to practice religion in words than it is to practice justice
in action. If we want to look at the Gospels with our eyes open, we will see
the truth. No one will get away with salvation by words. Only those who do
the will of Jesus' Father in heaven shall be judged worthy to join the kingdom

of heaven. So, if you see Jesus as I do, as a person who speaks from the highest authority, then you will want to know, as I do, exactly what is expected.

Jesus taught us to "Do unto others as you would have others do unto you." Since we all "hunger and thirst for justice," it is clear that what is expected from each of us is that we render justice to others. How we render justice to others is the way justice will be rendered to us. Again and again the mallet of metaphor and example thuds on our wooden heads. It were better that our heads be made of bronze. Then we might hear a bell, the sweet tone of cold truth.

The "grand finale" of how we shall be judged is presented in dramatic metaphor in Matthew 25, in the description of "The Last Judgment," which speaks eloquently to what is expected of us.

> The Last Judgment 31-40: (discussed also in Chapter Four)
> 31. "But when the Son of Man shall come in his majesty, and all the angels with him, then he will sit on the throne of his glory; 32. and before him will be gathered all the nations, and he will separate them one from another, as the shepherd separates the sheep from the goats; 33. and he will set the sheep on his right hand, but the goats on the left.
>
> 34. "Then the king will say to those on his right hand, 'Come, blessed of my Father, take possession of the kingdom prepared for you from the foundation of the world; 35. for I was hungry and you gave me to eat; I was thirsty and you gave me to drink; I was a stranger and you took me in; 36. naked and you covered me; sick and you visited me; I was in prison and you came to me.' 37. Then the just will answer him, saying, 'Lord, when did we see thee hungry, and feed

thee; or thirsty, and give thee drink? 38. And when did we see thee a stranger, and take thee in; or naked, and clothe thee? 30. Or when did we see thee sick, or in prison, and come to thee?' 40. And answering the king will say to them. 'Amen I say to you, as long as you did it for one of these, the least of my brethren, you did it for me.'"

I believe that what is said here in Matthew 25 is not the moral philosophy of a nice man, but the Law of the Universe. If we violate the law, we cannot argue ignorance of the law as our defense. We are each obligated to know what the law is. We could not have a civil society if a plea of ignorance were an effective defense. Anyone could avoid responsibility for virtually anything simply by pleading ignorance. You are informed. Reading this book reinforces the reality of being informed. Improving one's knowledge increases one's responsibility. That is apparently why many people choose ignorance, or even stupidity, on the mistaken premise that they will be forgiven if they say they did not know what was expected. Nowhere on Earth does human society forgive an offense because the offender claims he did not know the law. Why should the kingdom of heaven be any different? The kingdom of heaven resides in the same universe as we do. It is not a magical kingdom. It is the kingdom of life, logic and Natural Law. Nature does not listen to "closing arguments." There is no jury of twelve peers. If Matthew 25 describes the Law of the Universe, then there is a jury of One, and that jury is Nature. Regardless of whether or not you have a personal relationship with God, I want to assure you that God has a personal relationship with Nature. We are informed.

Chapter Seven: Three Unforgivable Sins

Here we are continuing our exploration of what kinds of behavior are expected from us, if we are to gain admission into the kingdom of heaven, the community of good servants. We know that Jesus tells us a great deal about what we should do. We naturally want to be as clear as possible about what we should not do. We see and hear Jesus being forgiving and telling us that the kingdom of heaven can be forgiving. However, we need to know if there are any behaviors, or behavior patterns, that will not be forgiven. I believe there are three. They are closely related, and in order of increasing seriousness, they are:

1) cynicism;
2) hypocrisy;
3) authoritarianism.

These three offenses, when established as patterns of behavior, will not be forgiven by the kingdom of heaven because of their pervasive damaging consequences to both children and adults, and their destructive impact on human development. These three patterns of behavior obstruct the opportunities for good stewardship. One could say that these three behaviors "close the gates of heaven." The three offenses, which are often committed simultaneously, are the three unforgivable sins, referenced in two passages in the Gospel, Matthew 12: 31-37, and Matthew 23: 1-39.

In Matthew 12 Jesus tells us we will not be forgiven if we "blaspheme against the Holy Spirit." In Matthew 23 Jesus tells us that the hypocritical and authoritarian behavior of religious and political authorities -- the Scribes and Pharisees in the time and place of Jesus -- will not be forgiven. Let's take a look.

Matthew 12: 31-37:

31. "Therefore I say to you, that every kind of sin and blasphemy shall be forgiven to men; but the blasphemy against the Spirit will not be forgiven. 32. And whoever speaks a word against the Son of Man, it shall be forgiven him; but whoever speaks against the Holy Spirit, it will not be forgiven him, either in this world or in the world to come. 33. Either make the tree good and its fruit good, or make the tree bad and its fruit bad; for by the fruit the tree is known. 34. You brood of vipers, how can you speak good things, when you are evil? For out of the abundance of the heart the mouth speaks. 35. The good man from his good treasure brings forth good things; and the evil man from his evil treasure brings forth evil things. 36. But I tell you, that of every idle word men speak, they shall give account on the day of judgment. 37. For by thy words thou wilt be justified, and by thy words thou wilt be condemned."

There are words and phrases in this passage that can confuse or frustrate any reader. I believe I have figured it out and can show that this passage is consistent with all of Christ's teachings. First, let's look at what appears to be three unrelated thoughts or ideas:

One: To "blaspheme" against the Holy Spirit shall not be forgiven.

Two: After all that Jesus has said about our being judged according to our conduct, here he is saying that we will be judged by "every idle word men speak," or by our words.

<u>Three</u>: Jesus is once again using the metaphor of a tree and its fruit as a metaphor for a person and the results of that person's behavior. Just as a tree is known by its fruit, such as an apple tree or a fig tree, a person is known by the results that they produce in the world. Whoever produces good things, is called a good person, and whoever produces bad things, is called a bad person. This is obviously true in human society, but this concept seems to be trivial, not complex and deep in a way that establishes some connection with the profound concept of the Holy Spirit.

(Referring to Two above) It turns out that the concept of the good person bearing good fruit is in fact not trivial at all, and it is connected directly to the Eighth Commandment: "Thou shalt not bear false witness against thy neighbor." This is a clear command, and it pertains to a specific type of lie. There are many avenues for deception, but this form of deception in particular, to say something false about the behavior of another person, is serious enough to be included in the Ten Commandments. This reality makes it one of our fundamental obligations to be truthful and very careful when describing the behavior of another person. This issue is so important it has been a matter of civil law for centuries. It is against the law to slander another person, to discredit their reputation. To say that another person lies or steals, or is immoral or violent, falsely, could cause others to refuse to do business with them, and to continue the "false witness" and exaggerate it or add to it. This could cause others in the community to stop talking to that person. It could cause the victim of one's gossip to lose friends, business, or occupation. In other words, "false witness" can be extremely destructive. This should be clear to everyone in our own times. If a man is convicted of molesting a child, and he is later found to be innocent based on new evidence, he is often unable to overcome the willingness of the community to believe the original story of his offensive behavior. Once a person's reputation is damaged, repair is difficult or impossible. Their life is changed, sometimes destroyed. Their opportunities to do good, to be good, to benefit society with their creativity, is gone, lost to society because of gossip. (Referring to Three above) Jesus also says here "make the tree good and its fruit good, or make the tree bad and its fruit bad... ." This is perhaps where a reader gets lost or distracted. Why is this passage here? In this passage, I see the definition of "cynicism," the first unforgivable sin. And we participate in this offense all the time.

> "Cynicism" means to be "cynical," and Webster's English Dictionary defines "cynical" as: "given to faultfinding...," "given to or affecting disbelief in commonly accepted human values and in humankind's sincerity of motive or rectitude of conduct..., accepting selfishness as the governing factor in human conduct..., exhibiting feelings ranging from distrustful disbelief to contemptuous and mocking disbelief. ... often implies a disbelief in sincerity, benevolence, rectitude, or competence."

One could say then that "cynicism" and being "cynical" means attributing no quality of altruism or benevolence in human conduct, but only selfishness. In other words, to practice cynicism means to view all behavior as self-serving, and never intended to accomplish any good for others or for human society. This is in reality a dark and bleak outlook when taken seriously, when the heart is frozen in refusal to believe that a human being can genuinely want good things for others. And this outlook is totally contrary to the first law: "You shall want the same good things for others that you want for yourself." This is the Islamic version of "The Golden Rule." To be cynical then, and to put the pattern of behavior called "cynicism" into practice, one would never believe that the Golden Rule is possible, never in effect. Everyone is acting only to get good things for themselves, and they do not want others to have those same good things. This is cynicism, and this is the first sin that is unforgivable, and it is obvious why it is unforgivable. It is unforgivable because if the cynical outlook were correct, the First Law of the Golden Rule would be impossible. Doing good is impossible, because all actions are selfish and only selfish. So, now we should be able to see why this behavior, though defended by some people as a form of "realism," is in fact the first unforgivable sin. It is first perhaps because it is the most common, and because it leads to the next two, which can be far more destructive.

Why do I say that Jesus is talking about "cynicism" here? Because his words describe the common pattern of cynical behavior that we designate as "gossip." Gossip is that form of communication, mouth to ear, that is customarily employed to spread negative opinions about other people. We already know that a negative opinion about another person can be a violation of the Eighth Commandment, if it is not based on direct knowledge of that person's behavior and is not true. Regulating this behavior legally is cumbersome and noxious. Many minor slanders go unnoticed or unpunished, but they are still a sin, and they are in fact the sin of cynicism. The pattern of behavior in our own times will be recognizable to everyone. We listen to voices and watch images on television that tell us who was caught misbehaving sexually, or who stole a car, or who was caught embezzling, or who was seen improperly dressed, who is not married but pregnant (that is a favorite), who is getting divorced after a short marriage, who is rich but also a mess emotionally, or drug addicted. We seem to thrive on bad news about people, especially people who have achieved some form of fame, such as singers or actors or artists or politicians. We also love to receive news about the troubles that have befallen our "enemies," or any other nation or society with whom we have some disagreement or issue. Again and again, whenever one of our "enemies" or favorite celebrities falls down, or is in trouble, we want to hear about it so that we can gloat over their misfortune. And -- here is the point Jesus is making -- whenever one of our enemies has done something that appears to be good (bears good fruit) such as a cure for a disease, a new technology, a program or policy that helps the common people, any good thing, we attribute their behavior to selfishness. When we do something good, it is because we are benevolent and democratic. When our enemy does something good, we attribute their "good fruit" to negative motives or selfishness. We say they have done something that only appears to be good, but in reality they have done it only to control people, only to dominate, only to advance their own hidden agenda, only because they want praise and to gain some advantage. We just cannot allow anyone to think that our enemies have done something good because they are good. We see the "good fruit," and we say that it comes from a "bad tree." That pattern of thinking brings us to the first concept - to blaspheme against the Holy Spirit.

Many times Jesus tells us that we are the same as he is. He is human, we possess the seeds of his powers within us. The Holy Spirit that is in him is in us, in all of us. This is the teaching of Jesus Christ, and that is why it is an unforgivable offense to make the fruit good and the tree bad, because that is blasphemy against the Holy Spirit. Blasphemy against the Holy Spirit is embodied in the act of refusing to see a positive motive in a person or society that we don't like. Blasphemy against the Holy Spirit grows out of cynicism, and is in fact the same offense as cynicism, which means refusal to acknowledge a positive motive -- the Holy Spirit -- in a person or society or community that we don't like, our "enemy." Therefore, Jesus is telling us that the other person, or other society or community is not our most dangerous enemy. Our most dangerous enemy is "cynicism." The other people who we don't like cannot hurt us as much as we can destroy ourselves with our cynicism. Read the passage from Matthew 12 again to see if you understand it differently from before. Decide for yourself whether it is about cynicism.

Next we look at the most important sociological document in the world, Matthew 23. In this incredibly important description of human behavior, the

Scribes and Pharisees are the religious and political authorities in the theocratic society of Roman-occupied Judea. Jesus' critique of how humans behave when vested with authority is universal. It applies to all human authorities in all societies and every level of human organization: family, club, tribe, town, city, church, state, business entity, educational institution, nation.

<u>Matthew 23: 1-39</u>:

1. Then Jesus spoke to the crowds and to his disciples, 2. saying, "The Scribes and the Pharisees have sat on the chair of Moses. 3. All things, therefore, that they command you, observe and do. But do not act according to their works; for they talk but do nothing. 4. And they bind together heavy and oppressive burdens, and lay them on men's shoulders; but not with one finger of their own do they choose to move them. 5. In fact, all their works they do in order to be seen by men; for they widen their phylacteries, and enlarge their tassels, 6. and love the first places at suppers and the front seats in the synagogues, 7. and greetings in the market place, and to be called by men 'Rabbi.' 8. But do not you be called 'Rabbi'; for one is your Master, and all you are brothers. 9. And call no one on Earth your father; for one is your Father, who is in heaven. 10. Neither be called masters; for one only is your Master, the Christ. 11. He who is greatest among you shall be your servant. 12. And whoever exalts himself shall be humbled, and whoever humbles himself shall be exalted.

13. "But woe to you, Scribes and Pharisees, hypocrites! because you shut the kingdom of heaven against men. For you yourselves do not go in, nor do you allow those going in to enter.

14. "Woe to you, Scribes and Pharisees, hypocrites!

because you devour the houses of widows, praying long prayers. For this you shall receive a greater judgment.

15. "Woe to you, Scribes and Pharisees, hypocrites! because you traverse sea and land to make one convert; and when he has become one, you make him twofold more a son of hell than yourselves.

16. "Woe to you, blind guides, who say, 'Whoever swears by the temple, it is nothing; but whoever swears by the gold of the temple, he is bound.' 17. You blind fools! for which is greater, the gold, or the temple which sanctifies the gold? 18. 'And whoever swears by the altar, it is nothing; but whoever swears by the gift that is upon it, he is bound.' 19. Blind ones! for which is greater, the gift, or the altar which sanctifies the gift? 20. Therefore, he who swears by the altar swears by it and by all things that are on it; 21. and he who swears by the temple swears by it and by him who dwells in it. 22. And he who swears by heaven swears by the throne of God, and by him who sits upon it.

23. "Woe to you, Scribes and Pharisees, hypocrites! because you pay tithes on mint and anise and cumin, and have left undone the weightier matters of the Law, right judgment and mercy and faith. These things you ought to have done, while not leaving the others undone. 24. Blind guides, who strain out the gnat and swallow the camel!

25. "Woe to you, Scribes and Pharisees, hypocrites! because you clean the outside of the cup and the dish, but within they are full of robbery and uncleanness. 26. Thou blind Pharisee! clean first the inside of the cup and of the dish, that the outside too may be clean.

27. "Woe to you, Scribes and Pharisees, hypocrites! because you are like whited sepulchres, which outwardly appear to

men beautiful, but within are full of dead men's bones and of all uncleanness. 28. So you also outwardly appear just to men, but within you are full of hypocrisy and iniquity.

29. "Woe to you, Scribes and Pharisees, hypocrites! you who build the sepulchres of the prophets, and adorn the tombs of the just, 30. and say, 'If we had lived in the days of our fathers, we would not have been their accomplices in the blood of the prophets.' 31. Thus you are witnesses against yourselves that you are the sons of those who killed the prophets.

32. "You also fill up the measure of your fathers. 33. Serpents, brood of vipers, how are you to escape the judgment of hell? 33. Therefore, behold, I send you prophets, and wise men, and scribes; and some of them you will put to death, and crucify, and some you will scourge in your synagogues, and persecute from town to town; 35. that upon you may come all the just blood that has been shed on the Earth, from the blood of Abel the just unto the blood of Zacharias the son of Barachias, whom you killed between the temple and the altar. 36. Amen I say to you, all these things will come upon this generation.

37. "Jerusalem, Jerusalem! thou who killest the prophets, and stonest those who are sent to thee! How often would I have gathered thy children together, as a hen gathers her young under her wings, but thou wouldst not! 38. Behold, your house is left to you desolate. 39. For I say to you, you shall not see me henceforth until you shall say, 'Blessed is he who comes in the name of the Lord!'"

The critique of the behavior of the Scribes and Pharisees is a critique of authoritarian behavior, whenever and wherever it occurs. It is obviously a critique of hypocrisy. To paraphrase what Matthew reports as Jesus' words in Chapter 23:

-- They assign hard work to others, but live a life of leisure themselves;

-- They put the symbols of good behavior on display in order "to be seen by men." The phrases "widen their phylacteries" and "enlarge their tassels" means they make it look like they have prayed more than they have.

-- What they value most is how they appear and their social status in the upper class, the "first places and front seats..."

-- They do not promote a true understanding of the kingdom of heaven.

-- They exploit vulnerable women, such as widows, and take their money as contributions.

-- They "convert" others to their corrupt way of protecting their selfish interests.

-- They demand precision in ritual and bureaucracy, but not in social justice. They waste time and money on the minutiae of ceremony and procedures, but leave "right judgment and mercy and faith" to chance. By their attention to formalities rather than substance, they "strain out the gnat and swallow the camel."

-- Again, they are obsessed with outer appearances in that they "clean the outside of the cup and the dish, but within they are full of robbery and uncleanness." They are pretentious, hypocrites, because the appearance they create of just behavior is false and in reality they are liars and thieves. This sounds so familiar, two thousand years later. This pattern of human behavior must be deeply engrained in the human species.

-- The authorities are like "whited sepulchres," which are stone tombs covered with a white wash of limestone. Thus, they only have a thin outer coating of whiteness, but inside they are the decay and stink of death.

-- Then Jesus spells it out as clear as a bell (Matthew 23): "So you also outwardly appear just to men, but within you are full of hypocrisy and iniquity." I see this as the definition of hypocrisy: to pretend to be honest, when lying; to pretend to be ethical, while cheating; to pretend to be just, while stealing and betraying others; and in particular presenting the facade of a strictly religious person by pretending to support and practice adherence to high standards of moral self-discipline when in fact you are committing every offense that you claim will condemn the soul to eternal loss.

-- The authorities belong to a long line of traditional rulers who uphold traditions of social control but kill anyone, such as the prophets, who are critical of their behavior and call for change in religious and political institutions.

Looking back at the beginning, what is the meaning of Jesus' statement "You shut the kingdom of heaven against men. For you yourselves do not go in, nor do you allow those going in to enter." This looks to me like the authorities controlling knowledge, limiting access to information and limiting communications among the people. This would be consistent with censorship, secrecy in government, publicly discrediting and discouraging competing ideas, opposing desirable political change by calling it "unpatriotic," government control of the press, obstructing the publication of challenging ideas, or ultimately the burning of books. Jesus also says here "Therefore, behold, I send you prophets, and wise men,..." And those sent are tortured and killed. Looks the same as the Spanish Inquisition, the execution of Giordano Bruno and many other Renaissance scientists. Looks like the ongoing execution of female herbalists over the centuries on the grounds that they were witches. Looks like the Roman Catholic attacks on the Cathars and the Knights Templar in Southern France, apparently an attempt to exterminate them. Looks like the Muslims "converting" millions of people by offering them the alternative of immediate death by the sword. Looks like Christopher Columbus and Hernando Cortez claiming ownership of a continent. Looks like American officers giving blankets with smallpox to Native Americans. Looks like Adolf Hitler and Joseph Stalin executing political opponents. Looks like the United States of America constantly fighting wars to "spread democracy" while continuing to exercise control over banking and natural resources throughout the world.

Looks like authoritarianism is not an aberration, but a human habit as common and compelling as eating or wearing clothing. And this is why I say that Matthew 23 is the most important document in the world. It is our portrait. This is us, not "them." This is the problem of the species that must be overcome before we can become good stewards and be worthy of membership in the kingdom of heaven. This is why my ears hear Jesus teaching us the standards of the kingdom of heaven, that much can be forgiven, even blaspheming against the Son of Man. However, there are three sins that will not be forgiven. We know what they are. We are informed. When caught in the act, we cannot say "I didn't know. I must be stupid." Voluntary stupidity is like a fourth unforgivable sin. It is equally destructive as the first three.

Chapter Eight: Jesus' Identify and Mission

For me, the most significant passage in which Jesus' identifies himself and his mission is Matthew 10: 16-23, where he is advising his disciples.

> 16. "Behold, I am sending you forth like sheep in the midst of wolves. Be therefore wise as serpents, and guileless as doves. 17. But beware of men; for they will deliver you up to councils, and scourge you in their synagogues [churches, mosques, courts, village squares, soccer stadiums], 18. and you will be brought before governors and kings for my sake, for a witness to them and to the Gentiles. 19. But when they deliver you up, do not be anxious how or what you are to speak; for what you are to speak will be given you in that hour. 20. For it is not you who are speaking, but the Spirit of your Father who speaks through you. 21. And brother will hand over brother to death, and the father his child; children will rise up against parents and put them to death. 22. And you will be hated by all for my name's sake; but he who has persevered to the end will be saved. 23. When they persecute you in one town, flee to another. ..."

I compare this passage with the historical information we have regarding moral philosophers such as Socrates, and religious teachers such as the Buddha. I have never heard of any other great teacher who so accurately predicted how humans would use and abuse the power of his teaching. This passage is an accurate description of the two thousand years of human behavior that has followed the mission of Christ. How did he know that this is what we humans would do in response to his life and teaching? The correct answer is that he is a social scientist with enormous qualifications. Not only has he been a student of human nature on Earth, he also possesses a body of knowledge about how technological animals (human-like beings) have behaved on other planets when they were sent teachers (prophets) to tell them what they need to know about how the universe works. There is no common understanding. Different individuals understand the substance of the teaching differently. Institutions establish doctrines and form their battle lines. Some claim Jesus' teaching as a moral philosophy worthy of study. Others claim Jesus' teaching as a set of rigid requirements for the regulation of every human activity, the blueprint for an authoritarian theocracy, which is exactly how the Roman aristocracy used it to create one of the world's most successful and violent theocratic states. Many simply interpret the teaching to convey what feels right to them, or as confirmation of their hatred and resentment toward others.

What this passage in the Gospel says to me is that Jesus predicted accurately that we humans would fight among ourselves more or less continuously, and employ incredibly sadistic violence, over what is the true meaning of his teaching and what is the right way to apply it in our own lives. This includes the behavior of Moslems as well as Christians. Look at the passage again. Jesus said that whoever understands my teaching, or "the will of my Father in heaven," and applies it properly, will be persecuted, betrayed by friends and family, killed. This has occurred in all parts of the world. What world is more violent than the Christian world? Where has Christian evangelism not been accompanied by massive bloodshed? Even the American Civil War was driven in part by claims as to what was Christian and what was not Christian. The defenders of the historically unique and extreme form of slavery practiced by the owners of Southern plantations could pretend to be Christian and offer a pretense of self-justification only by means of the bizarre claim that Africans were not human. The Spanish Conquistadores who represented Christian Europe with their invasion of South America greeted the native population by cutting off their hands and killing their children. However sublime the message of Christ, those delivering that message certainly have been lacking in social skills. The frustration with "Christian" nations can produce castles filled with books, but none of them, however detailed, trump the concise accuracy of Matthew 10: 16-23. Jesus, a man who was not only repetitive but also a man of few words, says what needs to be said succinctly. By studying Jesus' words, one can see that substance is not to be measured in pages. This passage is indeed only a few words, but it is the essence of the history of Western Civilization since his trial in Jerusalem. How did he know? Jesus also said that he did not come to destroy the old law (of Moses) but to uphold it, such as in Matthew 5: 17. He also is described as opening his heart and the teaching of the kingdom of heaven to all the world, not just the Jews, in a wonderful story of a wise woman with a sick daughter, in Matthew 15: 21-28. She was a Canaanite, not Jewish. The Canaanites once occupied the land of Israel and Judea.

21. And leaving there, Jesus retired to the district of Tyre and Sidon. 29. And behold, a Canaanite woman came out of that territory and cried out to him, saying, "Have pity on me, O Lord, Son of David! My daughter is sorely beset by a devil." [The daughter is mentally ill.] 23. He answered her not a word. And his disciples came up and besought him, saying, "Send her away for she is crying after us." 24. But he answered and said, "I was not sent except to the lost sheep of the house of Israel." 25. But she came and worshipped him, saying, "Lord, help me!" 26. He said in answer, "It is not fair to take the children's bread and to cast it to the dogs." 27. But she said, "Yes, Lord; for even the dogs eat of the crumbs that fall from their master's table." 28. Then Jesus answered and said to her, "O woman, great is thy faith! Let it be done to thee as thou wilt." And her daughter was healed from that moment.

To understand this incident correctly, we must remember that the Canaanites were sometimes the political and military adversaries of the Jews. Also, the Gospels tell us that Jesus taught his disciples how to heal physical and mental illnesses, which they supposedly did, though not as well as their teacher. This is why they say, "...she is crying after us." Jesus' disciples probably were possessive of him, thinking of him and talking about him as though he was strictly a prophet sent to the Jews. Others were deemed unworthy of the wisdom of a Jewish prophet, and certainly not eligible for healing by a Jewish prophet. Here Jesus affirms the universality of his identity and mission. It seems almost as though this woman could have been his accomplice in a plan he devised to correct the provincialism and self-centeredness of his disciples. He even says himself, at first, that he was sent only for Israel. When the woman pleads with him in such a down-to-earth and charming way, to say that she is only asking for a crumb from his table, just a crumb of his healing power, Jesus then dramatically affirms that he is for everyone and available to everyone who recognizes that his mission is universal. This is contrary to the theological interpretation of historians who argue that Jesus was a Jew speaking only to Jews.

The "Transfiguration." (Matthew 17: 1-9)
Here, we have a story emphasizing the spirituality of Jesus and the concept that he is from another world, and that the kingdom of heaven has sent him and communicates with him. What is especially powerful here is Jesus' response when Peter, who sees him "transfigured" suggests that Jesus' disciples set up tents to start a new religious organization.

> 1. Now after six days Jesus took Peter, James and his brother
> John, and led them up a high mountain by themselves. 2.
> and was transfigured before them. And his face shone as the
> sun, and his garments became white as snow. 3. And behold,
> there appeared to them Moses and Elias talking together

80

with him. 4. Then Peter addressed Jesus, saying, "Lord, it is good for us to be here. If thou wilt, let us set up three tents here, one for thee, one for Moses, and one for Elias." 5. As he was still speaking, behold, a bright cloud overshadowed them, and behold, a voice out of the cloud said, "This is my beloved Son, in whom I am well pleased, hear him." 6. And on hearing it the disciples fell on their faces and were exceedingly afraid. 7. And Jesus came near and touched them, and said to them, "Arise, and do not be afraid." 8. But lifting up their eyes, they saw no one but Jesus only. 9. And as they were coming down from the mountain, Jesus cautioned them, saying, "Tell the vision to no one, till the Son of Man has risen from the dead."

Setting up three tents and proclaiming that the tents were on the spot where Jesus was transfigured would have been application of a standard procedure for marking a place as holy, and later building a temple at that location, and later collecting money from pilgrims who visited the temple for support of the temple and the temple staff. The Greeks had done this, and the Persians and the Egyptians, and the Jews, and probably every ancient culture able to build a temple and impress the common people with some technical "magic," or answering a question or predicting the future. Temples were holy places, and both the poor and the rich paid good money to obtain the blessings of the gods associated with the temples and witness the magic of the priests, whatever that may have been. When Jesus says, "Tell the vision to no one," he brings their plan for making a living as temple priests to an end. If it is not to be identified as a holy place, then there is no basis for the construction of a temple, no basis for starting a new religious organization. This is not the only passage in the Gospels where Jesus implies that his mission is definitely not to establish another church. The disciples with him detected an opportunity for the good life as wealthy and influential priests living in comfort in a temple compound. The title of "Pope" and "Cardinal" had not yet been invented, but the general idea was not new. It is very interesting that this passage is "admitted" into "the canon," meaning it was approved as valid Christian Gospel as part of the agreement between the early Christians and the Emperor Constantine, so that the new Roman religion could become the officially approved, universal or "catholic" religion of the Roman Empire. It is ironic, because Jesus said, "No, don't put up tents, don't start a new temple or religious organization based on me." And this statement by Jesus is included in the Gospel that was adopted in the process of forming the new Roman Catholic Church, which was and is presented to the world as the church based upon him. In other words, if you see the Gospels as I do, the formation of the Roman Catholic Church is an act of betrayal of Christ's will. In any case, it is not Jesus' mission to establish a new religious institution. His mission is to correct the misunderstandings that we humans have about who we are, and what the kingdom of heaven is, and what is expected of us if we are to join the kingdom of heaven. Jesus teaches that we will be judged according to our conduct, not according to the organizations of which we are members. Jesus said, again in plain, concise language (Matthew 23) that loyalty to an organization is not and can never be the same thing as faith in our ability to understand reality, loyalty to truth and justice, freedom for all to find their path to the enlightenment or benevolent God they seek. These three universal ethical values:

1) faith in our ability to understand reality and in the goodness of human beings;

2) loyalty to truth and justice;

3) freedom for all to find their path to the enlightenment they seek;

are the greatest virtues an individual, a society or a civilization can practice, the opposites of the three unforgivable sins of cynicism, hypocrisy and authoritarianism.

Part II: A Traveler's Guide

Chapter Nine: Weeping and Gnashing of Teeth

From this Chapter Nine forward, we are no longer focused only on the true content of Jesus' message to us, but also the important issue of how can one apply the wisdom of Jesus' teaching to one's own life. We are each and all on a journey, a journey through time and space, the journey that is your lifetime. What do we do with a life time? What is the best way to go on this journey?

Before we go on a journey to places we have not been before, we really appreciate having someone tell us what to look for, what to expect, especially what are the dangers. Is there anything that I must keep in mind to be sure to do? Anything to be sure to not do? What is the worst thing that could happen on this journey? Jesus answered that question. He told us what is the worst thing that could happen to us on our life journey, one's life journey as an individual and our life journey as a species. We could lose everything. If we do not understand the value of what we have, and fail to provide proper care for what is entrusted to us, we could lose it all. That total loss, every opportunity gone forever, would cause us to weep and gnash our teeth, to hate ourselves for having made the worst of all irreversible errors. This means to not only suffer the total loss of relationship and belonging, but also to have that pain compounded by the certainty that this deepest of all losses was deserved. I find five occasions where Jesus says that after a negative judgment, after people are refused admission into the kingdom of heaven, "...there will be the weeping and the gnashing of teeth." Matthew 8:12 (The Centurion's Servant); Matthew 13:42 (The Parable of the Weeds); Matthew 22:13 (The Marriage Feast); Matthew 24:51 (The Need for a Good Servant to be Vigilant); Matthew 25:30 (The Parable of the Talents).

This reaction to total and final loss is, to me, the description of self-hatred and profound regret. It is the reaction of asking oneself, "How could I have been so stupid?" It is not sadness, not the agony of burning in hell. I do not believe there is any "place" that is hell. Hell may be only this deep and irreversible depression and rage against oneself because we had a chance for heaven and lost it forever.

We are now discussing what kind of behavior is required of us in order to earn admission into heaven, how to avoid the irreversible tragedy of "the weeping and the gnashing of teeth." In Matthew 5: 23-26, is the answer to any further questions as to what is required.

> 23. "Therefore, if thou art offering thy gift [to God] at the altar, and there rememberest that thy brother has anything against thee, 24. leave thy gift before the altar and go first to be reconciled to thy brother, and then come and offer thy gift. 25. Come to terms with thy opponent quickly while thou art with him on the way [to settle a dispute in court, to settle a dispute in life]; lest thy opponent deliver thee to the judge, and the judge to the officer, and thou be cast into prison. 26. Amen I say to thee, thou wilt not come out from it until thou hast paid the last penny."

This passage is entirely consistent with the story of "The Last Judgment" in Matthew 25, where Jesus tells us that wherever and whenever we nurtured a fellow human being who was in distress, or weak and vulnerable (due to hunger, thirst, lack of shelter, social isolation, illness, a criminal error in judgment or self-control), instead of exploiting that person and using them selfishly, it was as though we did the same for him, for Jesus or for God. In this passage in Matthew 5 Jesus is saying nothing less than that God is not interested in our gifts. Not interested in our cathedrals, our statuary, our song and music, our paintings and sermons. He is saying that our true beliefs, our true religion, the behavior that reveals whether or not we are good stewards is not in our relationship *with God*, however genuine or contrived, but in our relationship *with our fellow human beings.*

Social justice is the Gospel. God is not moved if we say we love God or think we love God or we give things to God. God is watching how we love our neighbor and that will be the basis of judgment according to our conduct. Our gifts on the altar are not what God wants, not what the world needs. Our behavior toward others shows God our religion and who we are.

The phrase, *if you remember that your brother has anything against you,* does not mean only a civil dispute in the restricted legal sense, such as owing someone compensation. That is just the metaphorical setting, the setting of a civil dispute in court. The symbolic setting of a dispute in *court* is the equivalent of the real setting of any *dispute or conflict or issue* we have with our neighbor in everyday *life*. The deeper meaning, which I believe is not as "hidden" as some people might wish it were, is that what "your brother has against you" is any injustice that was committed once or any injustice which is in fact an ongoing injustice that is built into a social system or the political structure of any human society: caste, slavery, economic roles assigned by race, racial hatred, racial violence, oppressive economic systems, aristocratic manipulation of banking and financial institutions in order to maintain a contrived social, political and economic hierarchy, cynicism, hypocrisy, and authoritarianism, built into the social and political system. The direct opposite of what Jesus teaches us is social and political violence, murder, assassination, and war.

I would be far from the first to observe that human civilization appears to be intractably violent. When we teach our children history, we teach them about wars and the military leaders who won and lost the "great" battles. We teach that those values that are truly good, such as freedom and democracy and plenty of food and clean water and public health are all won by violent war. The "real world" that is often depicted for us both as children and as adults, in books, in movies, on television, is a world of sociopathic liars and thieves, killers, and sadists who rise to positions of power and control the direction of society. Why is this? This is not just about those who are designated as destructive and evil by the West. A serious study of American history will reveal that many of our leaders, including those whom we revere, have been corrupt in the clearest sense of the term. They have lied and cheated in order to get elected. They have cooperated with organized crime and drug dealers. They have used the power of government for the most common form of corruption, assigning government contracts to businesses that overcharge or that in fact were less qualified to fill the contract fairly than others who were not corrupt. There are leaders who started wars for the financial gain of a privileged class, and who often executed secret plans to create the appearance that the United States was an innocent victim rather than the creator of the state of war for selfish gain. Several historians, including this author, have argued that America's wars have often been wars of aggression to obtain or maintain economic benefits, while the government tells the people the wars are "wars for democracy."

One such study is *The Savage Wars of Peace*, by Max Boot (Basic Books, 2002). America's efforts to "spread democracy" are open to criticism, unfortunately, as being more like opportunities for America to secure access to minerals, farmlands and cheap labor. The world hates a liar and a hypocrite. We convey to our citizens a constant stream of propaganda, even in video games played by children, that life is dominated by violent battle, and there can only be one winner and one loser. Where in the mass media do we find the depiction of mediation and reconciliation? Where do we find the depiction of social justice, political justice, the depiction of errors being corrected, of justice being restored, the depiction of ordinary human beings building and maintaining among themselves the just relationships and permanent systems of social and political justice *like that justice of good stewardship which is described in the Gospels by Jesus?* On the basis of what logic do we tell the world, and ourselves, that we are a Christian nation? I know that we hope to be a Christian nation, and that is why I wrote this book. How much change is required in order for us to be in compliance with the teaching of Jesus Christ in the New Testament Gospels?

The industrial west was the first to voraciously burn up the fossil fuels of the planet and change the atmosphere we breath. Climate change induced by careless practices is likely to increase the destructive forces of severe storms, droughts and floods. Our chemical society has produced widespread genetic damage in the world. Hormone disruptors from human-made chemicals are found in the fat of polar bears, seals and birds. This vast damage to nature's balance is described in *Our Stolen Future*, by Theo Colburn, Dianne Dumanoski and John Peterson Myers (Penguin Books, 1997). Some birds and reptiles have lost their capacity to mate and reproduce or care for their young. It has been reported that the Gulf of Mexico has a huge "dead zone" due to the chemical pollution that has flowed into the Gulf from the Mississippi and other rivers of the United States. It has been reported that a gigantic circle of rotating garbage, mostly plastic, can be seen in the Pacific Ocean by passing boats and by robotic satellites above. (See National Geographic, May 2005, National Oceanic and Atmospheric Administration, Microbial Life Education Resources, and many other sources.)

Numerous animals and plants are at risk of extinction. The fish and shellfish of the oceans, and the giant outgrowths of coral reefs are dying away due to the careless growth of human populations and careless use of chemicals that nearly always have "unintended consequences." One unintended consequence of global warming is that the acidity of the oceans has increased, which diminishes the capacity of all forms of shellfish to manufacture the calcium carbonate required for their shells. How long can we pretend that "unintended consequences" are acceptable. I am reminded of a verse in a song sung by Judy Canova that I heard as a child: "I didn't know the gun was loaded and I'm very very sorry, my friend." We do not know if the dead friend accepts the apology. Does our entire civilization present itself to the universe as one who destroys and then says "I didn't mean it." Why have so many dangerous chemicals, in such dangerous quantities, been developed and sold commercially at a profit only to have us discover later that these chemicals are dangerous to life? Are we innocent or stupid or irresponsible?

Our environmental errors appear to be gross violations of our commitment to good science. Scientists have known since ancient times the principle that any causal factor can have complex results in addition to the specific results desired. Even when doing something as simple as burning wood to create heat we also make smoke and soot. In the eighteenth and nineteenth centuries the burning of wood and coal caused widespread respiratory distress and illness. This form of atmospheric damage continues in those parts of the world where wood and coal are still primary fuels. The smoke from home fires and industrial coal furnaces hung in the air like dark curtains. The old solution was to change home fuels and have industrial plants build tall smokestacks so that the cough-inducing smoke and particulate pollution from industrial furnaces was carried higher into the atmosphere. When people nearby no longer smelled and choked on the industrial waste, they accepted their immediate perception that the problem had been solved. Any thinking scientist knew that we were still injecting chemical waste into the atmosphere, chemical waste which we later had to admit produced sulfuric acid, nitrous acid, carbon dioxide, heavy metals and other toxic compounds in the air that we (and all other living things) breath daily. Our wonderful insecticides and herbicides have been found to be deadly to birds and beneficial insects and beneficial bacteria and fungi in the soil. We just can't seem to get it through our heads that the fundamental principle of science is that there is no such thing as one cause that has one effect. There is no poison that poisons only one plant or one animal. All living things are related. Humans possess thousands of genes that are the same as genes found in plants. We need to awaken the intelligence of the spirit and admit to ourselves that as a civilization we look very much like we do not know what we are doing. We like to point to our music and art and poetry and architecture and positive technology, and innumerable "good deeds," but what good are these if we are killing the ecological system of life support on our planet?

A great deal of our science finds its way into military surveillance, military communications, lethal weapons and military logistics so that we can maintain armies to fight and kill and destroy. Even when we use our best chemical science we kill and destroy the life support systems that enable us to live on Earth. Do we, the people of Earth, still have any hope of rescuing the people of Earth from the people of Earth? Hope is not yet dead, I believe. Unless we have already been judged and live in the early stages of a psychological hell, where we suffer the anxiety of anticipating a future of decline, decay, and extinction, or we are simply cast backward in time into the animal world again, to live as dumb animals, having failed to live effectively as intelligent caretakers. I listen to the world and I fear that what I hear is the weeping. The weeping has begun. The teeth will soon be grinding, grinding down to blackened stubs. Are we standing over the edge of a cliff looking down, or are we already falling through empty space, wondering how far we have to fall? How soon will this falling end? How will it feel when its over?

This is what is meant by "the weeping and gnashing of teeth," the knowledge that by our own stupidity we have killed our family, destroyed our home, violated innocent children, murdered the Holy Spirit whom we failed to see in our neighbor. Our planet is dry and sick, cold yet burning with fever, the molested captive of a violent brute. Before we utterly destroy it, it is taken away from us and given to someone who knows how to take care of life. This is how it feels when the falling stops, when our body and soul experience the crushing collision with the solid and unmoving reality of Nature's way, the way eternal that has no name. Jesus told us in the parable of the vineyard that if those hired to care for the vines and harvest the grapes turn out to be thieves and murderers they will be destroyed and replaced. What part of "destroyed and replaced" do we not understand? Discussing "the weeping and gnashing of teeth" is painful. We hope for final forgiveness and approval. Jesus' description of ultimate failure and being "cast outside" is hard to accept. However, these passages about the harsh consequences of failure -- failure to understand the importance of good stewardship -- pull the warm blanket of love off of us on a frosty morning. The chilled air awakens us to the possibility that Jesus' message can be interpreted as deadly serious and icy cold. We must act on this information. We must not be paralyzed by fear. In Chapter Ten, we return to the brighter side.

I could apologize for the severity of "the weeping and gnashing of teeth," but if I did I would be apologizing for something Jesus said, which I certainly cannot and must not do. I am reporting it here because it needs to be heard and contemplated the same as the other parts of Jesus' message. I would like to emphasize once again that my understanding of Jesus' teaching is that the consequences of failure would not be the act of a person, would not be like being spanked by a father or mother, but is more like getting wet when it rains, cutting yourself if you are careless with a knife, getting burned while playing with fire. The harsh results of failure are an act of Nature, the natural consequences that occur when a technological animal fails to understand their crucial and inescapable obligation to take care of the natural world that is entrusted to them. Therefore, it is a key theme of this book that "the weeping and the gnashing of teeth" is a scientific reality about the natural world and is not a willful punishment imposed by an angry or violent anthropomorphic God. I do apologize for what may appear to be a grimly negative view of human behavior and human technology. However, we need to uphold the highest standards in our self-assessment. We must be precise and thorough in our evaluation of the risks we take by using our technology. Nature does not forgive or punish but only produces inescapable natural consequences.

Chapter Ten: Everything Will Be Revealed to Everyone

How will you be informed, on your lifetime journey, which way to turn? Jesus' life and teaching created many questions in the minds of his disciples, and mysteries as to who he was and by what authority he taught the people and cured illnesses, and told people that their sins were forgiven. Jesus told his disciples he would die and then rise from the dead. They wondered what that meant and if it meant exactly what Jesus said, why would that be the outcome of Jesus' life. What would be accomplished by that miracle? What would happen after Jesus returned?

> Matthew 10: (Where Jesus tells his disciples he is sending them out to heal and teach.)
> Then, having summoned his twelve disciples, he gave them power over unclean spirits, to cast them out, and to cure every kind of disease and infirmity. ... 16. "Behold, I am sending you forth like sheep in the midst of wolves. ... 26. Therefore do not be afraid of them. *For there is nothing concealed that will not be disclosed, and nothing hidden that will not be made known. ..."*

The priests and Scribes and Pharisees resented Jesus because he was not "certified." Jesus did not get his authority from the hierarchy of Jewish officials or from Rome. They were constantly trying to catch him breaking the law, or committing a serious offense against the religious rules and customs.

> Mark 11:
> 27. And they came back to Jerusalem [Jesus and his disciples]. And as he was walking in the temple, the chief priests and the Scribes and the elders came to him, 26. and

93

said to him, "By what authority dost thou do these things?" and, "Who gave thee this authority to do these things?"

Jesus then said that he would ask the priests a question first before he would answer their question regarding the source of his authority. He talked about John the Baptist and asked them whether the baptism practiced by John was "from heaven" or "from men." This meant asking whether John's "authority" was naturally spiritual, meaning "from heaven" or from an official human institution. They were afraid to respond either way, because if they said "from heaven," they would acknowledge their own guilt and rejection of John. And they knew that the people believed that John the Baptist was a real prophet, so neither did they want to respond that John's authority was "from men." So, they answered Jesus' question by saying, "We do not know." And then, because the authorities evaded the issue inherent in his question, Jesus responded in like manner to their question: "Neither do I tell you by what authority I do these things."

In the Gospel according to John (John the Evangelist, not John the Baptist), Jesus tells his disciples that the "Spirit of truth" will come to them and teach them.

> John 16:
> 12. "Many things yet I have to say to you, but you cannot bear them now. 13. But when he, the Spirit of truth, has come, he will teach you all the truth."

No matter how thoroughly anyone studies the Gospels, and no matter how certain anyone might feel about understanding the meaning of what Jesus said, there are still unanswered questions. When the disciples heard Jesus say that he would be gone for a time (after his execution by the authorities), but would return at some future date, they asked "When?" But Jesus could not tell them when "the end of the world" would come. He could only promise that he would always be with the people in spirit and would return at some distant future date. I believe that upon hearing this, his disciples may naturally have asked the most logical questions, such as: What will happen to us while you are gone? Will you protect us from harm? What will happen here on Earth while you are away? Is there anything special that we should do while you are gone? Is there something we can do so that you will come back to us sooner rather than later?

In response to these questions from his disciples, Jesus answered that there was nothing special they needed to do while he was gone, other than live in accordance with what he had taught them. He asked his disciples to remember him whenever they sat at a table to eat and "broke bread together." He also told them that what would happen on Earth while he was gone would be no different from any time in the past. There would still be the usual natural disasters from time to time, droughts, floods, storms, earthquakes, fires, and pestilence or plagues, wars and rumors of war. This is the origin of what I believe is the "myth" of the Apocalypse that we find in the New Testament. These scriptures have been taken to mean that there will be some special set of disasters that will come as part of the "end of the world." I believe this idea that there will be signs and natural disasters as part of "end times" is an error made by the disciples. All Jesus was saying is that neither he nor his Father in heaven would be controlling the natural course of events on Earth during his absence. Business as usual for our planet Earth means exactly what is listed in Apocalypse: droughts, floods, storms, earthquakes, fires, and pestilence or plagues, wars and rumors of war. We live on an interesting planet. The reason we are emotional may be because our planet -- if we think of our planet as a kind of organism -- is emotional. The ironic curse "May you live in interesting times" is paradoxical, because here on Earth all times are interesting.

Jesus' promise *"For there is nothing concealed that will not be disclosed, and nothing hidden that will not be made known,"* is the defining event of the "end times" or the "end of the world" as we know it, or better said, as "we don't know it." When nothing is hidden, when we finally know exactly what has happened on Earth and how and why, when we finally know exactly who and what we are, that will be the "end of the world." And this ending, where our ignorance and the "mystery" of our origin and destiny is replaced with full knowledge, the truth with nothing hidden from us, is the true end of the world. When nothing is concealed from us, we will be in a new world. We will then know from where Jesus got his authority. I already believe that he spoke from the highest authority, the authority of truth. There is no higher authority than the truth, and that is, to me, exactly what Jesus meant when he said, "Heaven and Earth will pass away, but my words will not pass away." (Matthew 24: 35)

We are also told by Jesus that he has already revealed an important truth to us:

> Matthew 13: 34-35:
> 34. All these things Jesus spoke to the crowds in parables, and without parables he did not speak to them; 35. that what was spoken through the prophet [from the Old Testament] might be fulfilled, I will open my mouth in parables, *I will utter things hidden since the foundation of the world.*
> Matthew 5: 18: "For amen I say to you, till heaven and earth pass away, not one jot or one tittle shall be lost from the Law till all things have been accomplished."

And in Matthew 25, the all-important chapter with The Parable of the Talents and The Last Judgment, Jesus says to the good stewards who nurtured the weak and vulnerable, *Come, blessed of my Father, take possession of the kingdom prepared for you from the foundation of the world;*

What is meant by these references *"since the foundation of the world,"* and *"from the foundation of the world"*? And what is meant by the statement, which Jesus made more than once, that *not one jot or one tittle shall be lost from the Law till all things have been accomplished* ? What are all those things that are to be accomplished? It all sounds like a plan, like a plan that was in place from the very beginning of life on our planet, the founding of the world. What then, is the purpose of Jesus' mission? What is hidden that he reveals to those who have ears to hear and eyes to see? Possibly our planet Earth is rare. Supporting life in abundance, Earth may be an "incubator" planet from which genes and organisms are harvested and used to spread life throughout the universe.

Or perhaps our planet is ordinary, and others just like it are found by the thousands in galaxies throughout the universe. Whether our planet is special, rare or ordinary, the issue for us is most likely still the question of whether we are good stewards or bad. No matter what questions or issues are mysterious or unresolved in the Gospels, the primacy of stewardship is clear. The kingdom of heaven needs good stewards to take care of life in every part of the universe where stewardship is possible. A good steward who is faithful over a few things gets promoted to exercise their proven trustworthiness and competence over many things, or many planets. Some things have been *hidden since the foundation of the world*, but *nothing will be hidden after all things are accomplished*. And *not one jot or tittle of **the Law** shall be omitted* as this destiny of ours and of life in the universe unfolds. This is what Jesus said. This is what Jesus taught. This is like the food at a church supper, both warm and cold. We get the warm and cold truth that is served by Nature at the church of the universe.

How will this happen? How will all the mysteries be solved? How will we come to understand the whole truth of who and what we are? By attending a church? Should we listen to scholars? Should we study ancient languages for years? Join a holy order and live in a monastery? Should we be sure that we listen only to those who have a doctoral degree or masters degree? In Theology? In History? In Archaeology? No, that is not what Jesus advised. Jesus advised us to listen to him, to hear his words and understand. And he advised us that we could understand, if we only had faith in ourselves, faith in our ability to understand. All we have to do is try this faith, try believing that the truth about life, about the universe, about heaven and earth is available to us, directly from Jesus, without an interpreter in between. Jesus never advised his disciples to create a church. Jesus said, and demonstrated, that ordinary people like you and me can understand on our own, just by listening to Jesus and looking at him, with ears that hear and eyes that see.

Matthew 11: 25-29:

25. At that time Jesus spoke and said, "I praise thee, Father, Lord of heaven and earth, that thou didst hide these things from the wise and prudent, and didst reveal them to little ones. 26. Yes, Father, for such was thy good pleasure. 27. All things have been delivered to me by my Father; and no one knows the Son except the Father; nor does anyone know the Father except the Son, and Him to whom the Son chooses to reveal him.

"Come to me, all you who labor and are burdened, and I will give you rest. 29. Take my yoke upon you, and learn from me, for I am meek and humble of heart; and you will find rest for your souls. 30. For my yoke is easy, and my burden light."

In this passage Jesus also implied that he had the assistance of heaven in completing his mission of teaching and healing the sick. He always spoke directly to the common people. He spoke directly to the people in the

Sermon on the Mount, and when he and his disciples fed the crowd with loaves and fishes, when he sat at the table with "sinners" (Matthew 9: 9-13), when he told his disciples to let children come to him and not to chase them away (Matthew 19: 13-15). If you want to know who Jesus was, and want to understand his message, do not ask scholars, who will tell you that it was all a "metaphor." Ask the Roman centurion; ask the Canaanite woman whose daughter was sick and dying; ask the woman with a hemorrhage; ask the man with a withered hand; ask the two blind men who were given their sight, ask John the Baptist; ask the mute demoniac; ask the possessed boy; ask the disciple Peter; ask Bartimeus, and Lazarus. Ask the unnamed multitude who were healed and enlightened by Jesus as he traveled about, and he "withdrew" when he knew the authorities were after him. None of these people who heard and saw Jesus were scholars. None of them were members of a holy order. None of them lived in a monastery or a palace or a castle or a university. Jesus spoke directly to ordinary people and he never told anyone to build an institution -- a church or a university -- between him and us.

Just look, and see. Just listen, and understand. Jesus' message is addressed to you, not to a priest, not to a scholar or a general or a warrior or a king or a legislature. If you want to understand the truth, why not trust Jesus, who said: *There is nothing concealed that will not be disclosed, and nothing hidden that will not be made known,* to you.

Chapter Eleven: It's a Dangerous World "Out There"

Yes, we live in a dangerous world. But I believe the danger is within the self first before it is found "out there" in the world. Jesus tells us that the greatest suffering occurs when the authoritarianism of the heart becomes the authoritarianism of the state. We often feel exposed and vulnerable. We fear that we will be deceived and robbed, that our property or our own children will be taken from us by evil people. If we were to make a list of the greatest dangers in the world, dangers not only to individuals but to society and civilization, what would be on that list?

Probably, violent enemies of our nation would be among the first to be listed, meaning people who want what we have, or who want to diminish or destroy our power, or who want to discredit our law, our values, our Constitution. Probably criminals would be on the list, and lying politicians who pretend to represent our interests while actually acting on behalf of a protected elite at our expense, using the authority of government for selfish purposes. Other great dangers include disease, the influence of drug dealers and organized crime. We might consider immoral behavior, and the promotion of immoral behavior for personal profit, such as selling pornography, among the greatest of dangers to society. In every culture, murder is a crime, and there are sexual offenses that are considered bad behavior but not criminal, and there are sexual offenses that are considered to be criminal, and sexual crimes that are actually prosecuted.

What if we asked "What is the greatest of all dangers in the world?" I believe the greatest danger in the world is the reality that our human civilization on Earth is schizophrenic (divided as two separate personalities). We act and speak as though we truly believe in two separate realities in the single universe. (The word "universe" means *one* action, word or thing.) We are mentally divided into the two separate and real worlds of the physical and the spiritual. Much of human literature, religious teaching, political dialogue, and philosophical thinking presents to virtually everyone the doctrine that there is a spiritual or moral reality that is distinctly separate from the physical reality of the real, physical universe. This schizophrenic division is most clearly expressed in the ongoing conflict, or discord, between the world of "science" and the world of "religion." Scientists have often spoken out to discredit religion, although if we examine such scientific arguments carefully, we will see that scientists most often consider themselves to be religious or to have a kind of faith, but they intend to discredit *religious institutions* and *religious doctrines* and the *authoritarianism* of religious institutions, not religion or faith itself.

Some religionists also strive to discredit science, although my perception is that they discredit themselves more than they discredit science. The proper way to bring science and religion together is to study both carefully and look for the areas of reconciliation, areas where science and religion agree, or can work toward rational reconciliation. The danger of the hostility between scientists and religionists, or evolutionists and creationists, or naturalists and creationists, is that the conflict rises to the level of a serious battle for political power, or control over the state or the nation. In the United States, this "battle" has recently been referred to as "culture wars." The battle taking place is not a "culture" war. It is a religious war, and it has started the same way religious wars of the past, wars of incredible violence, injustice and destruction, started. Such wars begin with the usual conflict between what is defended as science and what is defended as religion.

The Medieval Roman Church used to refer to a supportive member of the nobility as a "Defender of the Faith." Do we label anyone today as "Defender of the Science." And how do religious organizations define "Faith?" It appears to me that they define "Faith" narrowly and rigidly as their own Faith only and their definition of God only and as their institutional organization only. But I have argued from the beginning that the most fundamental principle of both science and religion is the same: that we can understand the universe as it really is. This is the core faith of science. Therefore, anyone who defends science earns the same complimentary title of "Defender of the Faith." Who defends science, defends the faith upon which it is based, the faith that we can understand so long as we do the work to find the truth that reveals the universe as it really is.

I see the schizophrenic division of science and religion, physical and spiritual, as sufficiently dangerous to cause the total destruction of human society, up to and including the extinction of the human species. The misuse of religion to support war, which is nothing less than murder authorized by the group (tribe, clan, state or nation), is the essence of the problem. Consider the Ten Commandments of Christianity, for example. The Fifth Commandment states "Thou shalt not kill," or in modern language "Do not kill anyone." But some Christian theologians argue that the original Fifth Commandment means "Do not commit murder." And they then go on to argue a great distinction between "murder" and "war," and some even argue that God tells us (through the Old Testament of the Bible) that there are times when it is right to kill other people.

I see this issue as completely transparent psychologically and sociologically. The state must have the "police power," meaning the power over human life, in order to be effective as a provider of law and order. However, what we do, and I mean all of us humans in every culture, is we treat murder as a crime and war as heroic. What is the difference? Well, many people both high and low defend war on many grounds that appear logical at first: We have to protect our territory, and our property, and our rights, and our lives, and our values, and our interests, and on and on. But I see the issue differently. The real difference between murder and war is that murder is killing authorized by an individual and war is killing authorized by the group (tribe, state or nation). The killing authorized by the individual is called "murder," and the killing authorized by the group is called "patriotism." Take a look again. The word "patriotism" means "loving your father." It may be argued that the "Patrium" means the "Father Land," but the land that is the Land of "Father" means the land that belongs to your father, no matter how you paint it. So, you are either defending the land of your father (because it will be yours later), or you are taking land for your father (and yourself). The manipulative message of "patriotism" is that if you really love your father, you will go kill your father's enemies, which is in fact the true phenomenon of war. The young do not start wars. They are busy attending to other interests. Wars are started by old men. Wars are started by those who want to be in control by means of acquisition, by means of ownership. Wars are started by those who want to take possession of something and exercise power over others.

The most commonly known concept of killing allowed by God is the concept of the just war, known also as the "just war doctrine." The idea that a war can be just, if governed by certain rational and moral principles and rules, is defended by the Roman Catholic church and other Christian denominations, and philosophical organizations, in books and on the Internet. However, an examination of the various expositions of the "just war" philosophy, whether in summary or in detail, with or without historical analysis, always amounts to the same result. The inescapable result of the "just war" argument is that it makes the church a political ally of the state. In particular, it makes the church an ally in support of those state authorities who need an armed force to pursue their goals, regardless of whether the state's goals can be proven to be truly just, or just truly selfish, and possibly just plain evil. Any authority claiming with any measure of legitimacy to represent a social or cultural or religious group can claim to have suffered a great injustice, or to be continuing to suffer a great injustice, which injustice can be remedied only by war, that is by killing the enemy. The "just war" doctrine advises us that such an authority, especially the government of a state or nation, can declare the necessity of warfare, and the citizen is then expected to be compliant and participate as a combatant when asked or told to do so by that official authority. This viewpoint is nothing more than a continuation of the deal made by the bishops of the early Christian church with the Emperor Constantine: Make us the religion of your state, and we will be your Christian army. That Christian army has been available to the state ever since. The circular debate goes on throughout history: Is the state an arm of the church or is the church an arm of the state? A deeper level of hypocrisy comes to the surface when we consider that the "just war" concept includes the argument that the "just warrior" must fight against combatants only, other soldiers only, and must not harm civilians. People need to think about what is being said here, and use the imagination to get the picture of war that is being painted. The implication that war transforms itself from unspeakable violence and injustice to a "moral war" if the teams of combatants fight only one another, attack only other combatants, kill only the opposition combatants, blow up, burn and maim only the adversary military organization, and uses only the amount of violence necessary to achieve a goal of peace and safety, then the war fits the definition of a just war. This effort to rationalize the murder that we call warfare makes war into a kind of cataclysmic sports event. It becomes tragicomically ridiculous. If two nations were both genuinely committed to this concept of war, we could transport both armies to some remote location in a large desert, have them fight it out, declare the nation of the last man standing to be the winner, and then have that nation define the new peace, and of course write the next chapter of history. No civilians would then be killed, not even inconvenienced, except for the tax burdens. This would be the model "just war," a kind of super-bowl with bullets, cannons, rockets and fire bombs, tanks, planes, ships and submarines. It is a lot more than shoulder pads and helmets, but it is just that the game of war requires more equipment than other sports. Add the hospitals, and the factories required to produce the weapons, ammunition, uniforms, supplies, fuel, vehicles, and on and on. It is a very expensive sport, this "just war" where combatants fight only combatants, for only reasonable goals allowed by the rules, and use only the force necessary to achieve a peace that is better than the peace that would have been realized without war. I am a social worker, and actually a psychologist in terms of the total content of my education and experience. This "doctrine" of a just war is the most artistically refined and self-serving total nonsense that the human brain can produce in order to make the sour and sickly poison of denial go down the throat. The defenders of war have no serious grounds upon which to re-define murder as a recommendation from the mind of God simply because it is authorized by an official human authority. That would make the officials of government the servants of God, a kind of inevitable theocracy. All the arguments in defense of war appear to me to be bent toward the purpose of attributing a higher purpose to the lowest of purposes, the purpose to dominate, acquire, humiliate. There is no God of War other than us.

My perception is that the God that forbids killing authorized by the individual could not possibly support the idea that killing authorized by a group is different, no matter how large or "legitimate" the authorizing group. By arguing that God says it is okay for the state to go to war and kill others makes God into a politician. This is not a compliment. If this is the alleged Creator God, the God who created the universe and deems life to be sacred, there is no way one can twist logic to defend murder because a group of citizens, rather than an individual, feels that it is the right thing to do.

There seems to be a common perception that if I believe killing is wrong then I must be a pacifist, or against war, or against a particular war. This does not logically follow from my position. What does logically follow from my position is simply that war is authorized by humans and only by humans. God never authorizes war, because war is killing and killing is murder. So, I say to you that you can have all the wars you want, but please do me, and

God, the courtesy of taking responsibility for your decision, and please stop making me ashamed of the human species by continuously claiming that your atrocities are authorized by God. To serve their own selfish purposes, people make their anger and hatred God's anger and hatred. But God is not angry. God is distraught, because her children act crazy.

The way into a new age worthy of the adjective "new" is through the reconciliation of science and religion. There has to be some kind of arrangement or peace treaty. We have to agree that both science and religion are best defined as efforts to understand the universe as it really is. Religionists have to stop using God to promote war. Scientists have to stop using science to discredit religion, but we are all still free to discredit institutional behavior (as it applies to all institutions) and the pattern of authoritarianism that is found wherever a human nose breathes. We all have to stop using God as the authority behind our actions. The human interactions that occur on Earth occur because they are what we want. If there is a war on Earth, it is caused by human desires and human decisions. Therefore, we can really begin a new age only with this kind of change in accountability. We must agree that there is one universe and whatever is moral or spiritual exists here, with us, in this same physical universe where we use the scientific method to learn and understand. We also need to use the scientific method to learn more about religion. That is what I endeavor to elucidate in Chapter Twelve.

Chapter Twelve: Learning About Religion

While on your lifetime journey, what can you do to be reconciled to your brother, and sister? You can learn about religion.

In my country, the United States of America, sometimes called "the land of the free," there are citizens claiming that we are not free because the Supreme Court has ruled that children cannot pray in the public schools. Firstly, the prohibition against prayer is not absolute. There is no doubt in my mind that there are conditions where praying in school, such as a classroom full of students praying together for a classmate who died, would not be deemed "unconstitutional." It is the regular institutionalization of sectarian prayer, such as reciting a *Christian* prayer at the start of every school day, that is unavoidably a promotion of religion and a promotion of a particular religion to the exclusion of other religious beliefs. Praying at the start of a school sponsored sports event is the same type of institutionalized prayer. This type of action is clearly a violation of the First Amendment and the fundamental democratic principle of the separation of church and state.

The First Amendment to the Constitution of the United States of America:
Congress shall make no law respecting an establishment of religion, or prohibiting the free exercise thereof; or abridging the freedom of speech, or of the press; or the right of the people peaceably to assemble, and to petition the Government for a redress of grievances.

I also agree completely with the Supreme Court in their decision that it is "constitutional" for children to *learn about religion* in a public school. I have been interested in that concept of "learning about religion" all of my life. I learned about religion myself by seeking books for beginners about religions other than my own Christian tradition. Then I moved on to books for those who are no longer beginners. I studied what are called the *world religions* or the *great religions*:   Buddhism, Confucianism, Christianity, Hinduism,

Islam, Judaism, Sufism, Taoism (in alphabetical order here) and the Celtic religion. I also have studied Native American cultures and as a student of political science I have studied Communism, which can be regarded as a religion. Communism, just as is true for any religion, is ethical and benevolent in both form and content when it is not corrupted.

While I have taken these personal studies seriously, I have spread myself thin, and I do not claim to be an expert in any religion other than my own Christianity. And I do not claim to be a scholar in the field called "Comparative Religion." To me, that field becomes bogged down in knowledge of minute facts about other religions while neglecting the enlightening truths about religion itself. My approach was not so much comparative as analytical. Many years ago I developed a systematic approach to learning about religion. I call my approach or method *The Seven Pillars of Religion*. I consider this approach to be an excellent way for children or adults to use the scientific method to learn what religion is and how it influences our lives, our behavior and our politics. I recommend *The Seven Pillars of Religion* to you as the best way to learn about religion, either publicly or privately, and as the best way for our civilization to prepare for the reconciliation of science and religion. The reconciliation of science and religion is a task that we must complete in order to avoid self-destruction and move ahead in our development toward becoming a species of intelligent being and good stewards. If we continue to engage in violent conflict attributed to science and religion, we will perish.

In brief, the seven pillars of religion are:
People, Calendar, Ritual, History, Teleology, Ethics, Institutions.
The seven pillars of religion are a set of seven elements that apply to every religion to some degree or another. Each "pillar" is a phenomenon that supports the religion and keeps it alive. Each "pillar" is a quality or characteristic of the religion, and for each religion, one of the seven pillars

might be more important or more influential than the others. You will see how these supporting pillars work together, and how one might be more important to some members of the religious organization than to other members. The supporting pillars can be studied together, but it is very helpful to study them separately. By studying each phenomenon separately, your study becomes more scientific and more objective. A scientific approach allows you to be the detached student of religion rather than thinking and feeling that because you are learning about another religion you are becoming a "convert" or betraying the religious faith that you inherited from your family. Learning about religion means exactly that; it does not mean changing your religious tradition or membership. It can change your beliefs and deepen your faith. It does mean changing your mind, changing your mind from ignorant about religion to informed about religion. It means being educated instead of insular, provincial, prejudiced, and possibly dangerous to life and peace on Earth.

In order to take a closer look at the seven pillars of religion, let's first look at a list of religions and put them into four categories, just to help us move forward with a scientific approach. The religions listed here could be shown on a circle rather than a vertical list. This is just a list for the purposes of this book. My apologies if any part of my approach offends or appears to categorize a religion, or religious faith, in a manner that anyone finds incorrect or offensive.

Expressing one's opinion on any issue of religion is a risky business. Please give me the benefit of your understanding. I am still a student of religion prepared to be informed.

A list of religions (Naturalistic, World Religions, Minor Religions, Tribal Religions)

Naturalistic and Tribal Religions:
Animism, Naturalism, Paganism, Mysticism
Deism, Materialism, Humanism, Science, Communism

The Great or World Religions:
Buddhism (Zen), Confucianism , Christianity, Hinduism (and Jainism), Islam (Sufism), Judaism, Taoism

"Minor" religions:
Bahai Faith, Shinto, Zoroastrianism, many others

Notes on this list of religions (Animism, Naturalism, Paganism, Mysticism):
By "Animism" I mean any set of beliefs based on the concept that there is an "animus" or spirit or soul that inhabits everything, or nearly everything, certainly everything that is alive. By "Naturalism" I refer to all forms of Nature worship. "Mysticism" has a long history as a philosophical category, an attitude perhaps that the real physical universe is intractably mysterious. The word "mysticism" has a specific meaning in literature. However, I include "Mysticism" here simply as a general belief system similar in content to a world inhabited by "spirits" or spiritual or mystical forces. "Paganism" is historically, in Europe at least, the form of Nature worship and Animism that came closest to being institutionalized and political. In the form of the Celtic religion, and perhaps Druidism, it was institutionalized. The Celts of Europe and Berbers of North Africa resisted the Roman Empire and later invasions of European culture more or less successfully. Though changed by historical experience, the Celtic religion is still alive in Ireland and Wales and exerts some influence throughout the world. The Christian calendar incorporates elements of the Celtic calendar. The Christmas tree is a pagan tradition. The

Berbers have not been conquered intellectually or physically. Historians have observed interesting similarities among the Iberians and Basques of old Spain, the Celts and the Berbers. They all appear to sustain traditions that have their origins in pre-historic times. They recognize the sacred in the female as well as the male. They reject the idea that an ordinary individual requires the services of a priest or an institution in order to engage in a relationship with the divine or with any form of the sacred or spiritual that a human being may seek or encounter. This first group of naturalistic religions emphasizes the emotional and spiritual experience of the sacred and the awesome beauty of reality in every day of life.

Deism, Materialism, Humanism, Science, Communism are all highly intellectualized versions of "natural religion." These are the most influential and powerful examples of human beings organizing and articulating their scientific beliefs about human society *in Nature* and how we can best understand both our internal reality (the mind, psychological science) and the external reality (physics). This second group of naturalistic religions adds systematic thought and knowledge, social and political and economic philosophy. They add the challenge of knowing, understanding and acting accordingly in the world to the emotional and spiritual experience, expressing a kind of cosmic desire that what is felt morally shall be known intellectually and put into practice. They propose or even demand that the most spiritual and sublime realities of moral existence must be enacted in the physical world, that religion is embodied in practice and action and social organization, systems of law and justice. In particular, in this second group religion is social and political in both form and content, and not expressed or experienced only as a bond with Nature or the divinity perceived in Nature. In the case of Communism, the moral imperatives of the spiritual and the divine are deemed to be honored only if they are practiced diligently in the economic system of the society, as well as in social and political relationships.

These two groups of naturalistic religions are joined together. The Celts possessed substantial knowledge about the physical world. The intellectual religious viewpoint invites and supports spirituality and a genuine reverence for the elegance of Mother Nature.

Listed again alphabetically: Buddhism (Zen), Confucianism, Christianity, Hinduism, Islam (Sufism), Judaism, Taoism. These "Great" or "World" religions possess the distinction that they, with the exception of Judaism, have commanded the largest nominal and actual memberships, counted today in the hundreds of millions. They are all, including Judaism, political. They all have a country and culture of origin that has played a major role in the history of human civilization. Indeed, it is more or less obvious that each of these great religions can be deemed to be a "civilization" onto itself, such as "Buddhist Civilization," or "Islamic Civilization" and so forth. I list Judaism and include "Judaic Civilization" for the obvious reason that Judaism, or the original "Abrahamic" religion is the pre-cursor of modern Judaism, Christianity and Islam. The population of Judaism is relatively small because traditionally one is Jewish by birth, although conversion to Judaism is possible, and because the population of Jews has been reduced by the Holocaust that accompanied the rise of a religious cult in Europe labeled as "Fascism." The "World Religions" are inherently political religions. Their members and their institutions have endeavored more or less continuously to control the political process one way or another and often by use of violent force. But there are two very important observations to be made here that any student of religion should keep in mind.

First: Rather than religious institutions controlling the political process, religion has been abused and violated by politicians and warriors who use religion for their own purposes, especially for the prosecution of war. Raising an army is always hard work, and risky, and expensive. The argument that subduing a "sub-human" enemy for the sake of God is a moral imperative, and making killing the enemy a "sacred duty," has been the most common advertisement used to sell war since history has been recorded. Only people oppress other people and they often do so by using religion as the weapon to discredit, demean and destroy the political opposition.

<u>Second</u>: Buddhism is substantially less political over time than the Abrahamic religions of the West. Judaism, Christianity, and Islam all appear to begin with violence and sustain it throughout their histories. They are all historically so entwined with politics and the formation of theocratic governments that their destructively violent histories may be the cause of concern that they are in decline. One might imagine detached Buddhist monks meditating while Christians and Muslims hack themselves to pieces. While the "western" religions are the focus of human consciousness today, it may be that they are of such great concern to us because we sense that we can no longer sustain the violence of the conflicts they cause. It may be that the religions of Western Civilization, which brought us centuries of war in Medieval Europe, World War I and World War II, the Cold War and currently the spectacle of Islamic Jihad and Al Qaeda, and a form of sexualized Christian fundamentalism in the United States, will be exposed as a kind of sickness cured only by rest. In any case, the intellectual content of Buddhism, Zen Buddhism, and to some extent also Confucianism and Taoism have been labeled as "Eastern Psychology." And they are just that. While the "Eastern" religions are not totally innocent of political influence, they have been historically focused on the self-reform of the individual. They serve society as a guide to why one is unhappy, and why humans often do what is immoral because they do what is impractical. Scholars and theologians in the West sometimes refer to Buddhism as a "godless religion," and this may be taken as an insult or an intent to discredit Buddhism. This particular observation or criticism of Buddhism, that it does not prescribe a belief in a defined Creator God, is based on the assumption by many westerners that a real religion requires a creator "God." The truth is, the

absence of a western-style "Creator God" in Buddhism can be taken as an enormous compliment. This reality tells us that if you pursue enlightenment, as was recommended by Gautama Buddha, and you practice meditation or Yoga and meditation, and if you "inquire within" in order to understand your true self and your place in the universe, you will be free to accept responsibility for what happens in your life and in the world precisely because you do not have the mysterious will of an anthropomorphic (human-like) God to blame for what happens and what does not happen in the world and in your life. Buddhism emphasizes "harmony" rather than "control." This means that when you are thinking and feeling according to Buddhism, you observe the universe in an effort to see it as it really is, and see your self as you really are, and you endeavor to keep your self in harmony with the forces of Nature that cannot be changed. This is different from Christianity and Islam, which seem obsessed with "converting" the world to their version of reality, where "evangelism" is a duty. Institutional membership in Christianity or Islam appears to require a commitment to control the world and control the minds of other people. This certainly does not appear to be compatible with a world that seeks freedom and the joy of life. Because of their habits of violence, theocracy and war, Christians and Moslems betray the God they claim as theirs, and are in opposition to peace on Earth and good will toward humanity. The western religions must separate themselves from politics, or they risk being rejected and shut down. Although Buddhism belongs in the list of World Religions, Buddhism does not have the same blood on its hands as do Christianity and Islam.

The Seven Pillars of Religion:

1)   People: the people who first practiced a religion, geography and demography, regions of dominance or influence, holy places, cultural and political life intertwined with religion. The peoples who later adopted the religion.

116

2) Calendar: marking of natural cycles and important historical events, solemn holy days, festive days and customs which involve symbols of religious ideas or religious history. The calendar can be of immense importance in any religion.

3) Ritual: marking important community events, seasons, special events; the points of change in the life of an individual: birth, adolescence, marriage, divorce, special achievements, death, symbols associated with each ritual or which may be the signs which announce or denote the ritual. Customs, practices or organized events described as "traditions" or holidays or holy days or traditional celebrations.

4) History: the birth and development of the religion, its founder or founders, the set of beliefs which define the religion, reformers, growth and change in the religion and its persistence over time; divisions, debates, disputes, the evolution of ethical ideas and religious doctrines through the process of history; the religious responses to changes in human technology and science; traditional beliefs and reformed beliefs adapted or changed because of new knowledge developed by modern science.

5) Teleology: cosmology, philosophy, ideology, concepts included in one of the five major areas of explanation of the human identity:

    A.   The identity, description and care of holy scripture;

    B.   The identity or description of a God, gods or Creator God;

    C.   The origin of life and the physical universe;

    D.   The meaning of life, or the identity of the human species and its purpose or destiny;

    E.   The meaning of death.

6) Ethics: social and sexual rules, regulations and customs; descriptions of good and bad behavior, or of moral versus immoral behavior; suggested or prescribed rewards or punishments, shame, guilt, retribution, atonement, forgiveness or mercy; social justice; being a just person; offenses against other individuals, against society, or against God or God's will.

7) Institutions: brotherhoods, sisterhoods, holy orders, church buildings and church organization's, temples, schools, monasteries, retreats, charitable or community service organizations; church hierarchies; councils or synods to establish doctrines or beliefs, seminaries or training schools, colleges and universities that provide for religious scholarship.

I propose the following steps be taken in support of learning about religion and to end the misuse of religion in support of war and every form of violence.

Initial Steps Toward the Reconciliation of Science and Religion:

1) Scientists and theologians agree that regardless of the validity or accuracy of our understanding of evolution, good stewardship is probably both scientifically and morally the highest foreseeable level of development of the human species.

2) Scientists and theologians agree that both physical and spiritual realities must exist in one universe or one existence that encompasses all things. There are no separate realities. Any observation suggesting separate realities must be a misunderstanding or misperception of the unitary reality that encompasses all occurrences and events. Whatever is spiritual is not magical and neither separate nor contrary to the physical reality explored with the tools of science.

3) "Science" and "Communism" should each be considered a kind of religion. Therefore, it is helpful to study them using the format of the seven pillars of religion. No great contrivance is required to acknowledge that "science" can be deemed to be supported by each of the seven pillars: *People, Calendar, Ritual, History, Teleology, Ethics, Institutions.* Communism in its original form, and prior to being politically corrupted, was and is both a religion and the most common form of human organization prior to the rise of cities and institutions of government. A high percentage of primitive and tribal societies treated land and water resources as communal, and never privately owned. And this practice has continued into modern times. Tribal societies are often totally confused by the modern concept of private property when applied to land or other natural resources. The sciences of anthropology, psychology, sociology and genetics all converge on the observation that human beings evolve as a community. The individual human being is extremely vulnerable for the first three to five years of its life, and totally helpless for about the first year. Feral children lack the most important marks of a civilized human. They lack language skills and computational and social skills. They are usually autistic or semi-autistic and have no developed sense of social obligation or emotional attachment to another human being. They display only physical needs. The story of "Tarzan," an infant lost in an African jungle and raised by gorillas to become a jungle-wise human is totally impossible. A human infant can become human only if raised by humans. This is proven fact. Since it is proven fact that we evolve "communally," it therefore follows logically that "communism" is the valid concept that the community is the true unit of human organization, the living unit that originally possesses the traits that enable survival on Earth. Towns, cities, states and nations are comprised of communities. In the case of extreme disasters and emergencies, we revert to communal needs, communal interdependence, and communal loyalties. The "gang" is the social unit that an individual will defend and die for, not the nation.

It is important that we accept the reality that Jesus supported communal organization. Many books have been written, many by Biblical scholars, citing historical evidence that Jesus lived in an intentional community such as that described in the Dead Sea Scrolls as the Essene community. Jesus is compared to a person described in the Dead Sea Scrolls as the "Righteous Teacher." Such communal organizations are referenced in ancient scriptures that were not available when the Romans and early Christians decided which books would be included in the Bible, or the "canon." These include the *Nag Hammadi Library*, *The Gospel of Thomas*, *The Gospel of Q*, and *The Book of Enoch* as well as many others. The four synoptic Gospels of Matthew, Mark, Luke and John imply that Jesus lived in an ethics-based communal society, where community members who were not related to him by blood were his "brethren, brother and sister." This is not understood by historians as an aberration or a suspicious cult, because in Jesus' time and place many groups of people sought what we might simply call "the right way to live," just as the utopians of modern times sought the right way to live in America - and many other nations - in carefully organized ideal communities.

It is extremely important that we clearly reject as "communism" the authoritarian and cruel form of government organization that was developed in the twentieth century by the Soviet Union, first briefly by Nicolai Lenin and for another twenty-nine years by Josef Stalin. The modern philosophers of communist organization, including Marx and Engels, argued that communism vested the greatest power in the common people, the working people, or the "proletariat." For most advocates of communism, a communist society was utopian, an ideal form of altruistic communal society that rejected materialism and any exaggerated economic hierarchy. There would be no "privileged class." The communities that came close to being communist in both spirit and practice would include the ancient communal organizations of Jesus' time, and the modern communities of Shakers, Amish, Hutterites and the caretaker industrial communities of utopians such as Robert Owen or even the nineteenth-century textile mills of Lowell, Massachusetts. The core value of all of these communal systems, old or new, is to carefully plan the social rules and structures toward the purpose that *the community be safe, nurturing and supportive of the physical, mental and spiritual development of the people living in it*. Hundreds, if not thousands, of advocates of altruistic communism existed long before the Russians tried to "jump" from a rigidly feudal society to a super-modern industrial communist state. The defenders of true communism argued that communism would develop *naturally* in societies that first made the transitions from agricultural or feudal society to republic, and then to democracy. Only after a highly developed democracy was established would the common people be able to exercise the power of the majority and implement communist policies as advanced refinements of a truly democratic society. This is the idea of *Communism* that is best understood when studied as a religion, with *People, Calendar, Ritual, History, Teleology, Ethics, Institutions*. A scientific study of Communism would include examination of its theft and corruption by an authoritarian state government. This would make the Soviet Union a kind of theocracy. A study of the Soviet Union will show that the ruling class actually imposed designated doctrines as forced "scientific" doctrines that could not be effectively challenged by real science. This pattern of behavior was more or less identical to that of the Roman Catholic Church in Medieval Europe when it imposed theological doctrines as fixed and final truth beyond question or re-examination. The original idea of Communism is neutral toward economic systems, and could employ free market principles, although it does impose an extreme level of economic equality. Therefore, the exaggerated economic hierarchy (inherited from feudal society) associated with western capitalism would be incompatible with Communism.

4) A "Peace Treaty" among religions:

As a crucial step toward the reconciliation of science and religion, I propose the following peace treaty among all religions. This peace treaty is not "pacifist." No teacher of religion or practitioner of any religion is required by this oath and treaty to denounce violence or refuse to participate in violence. The oath and treaty requires only that hatred, spiritual exclusion and violence not be attributed to the will of God.

Proposed Peace Treaty among all religions, to be subscribed to by religion teachers:

I am a teacher of religion and I acknowledge and actively support peace among all people and among all religions and religious institutions. I acknowledge and teach that there is one sacred universe of reality, one origin of life and one sacred love of life that inspires all living beings. I accept and teach that any and all acts of hatred or spiritual exclusion or violence are the sole responsibility of the human beings who commit such acts, and neither hatred nor spiritual exclusion nor violence can be an attribute of the Creator God. There can be no instruction or command to hate or kill given by the Sacred Spirit of Life that inhabits the one universe and every human being. Therefore, I shall never attribute human violence to God or the will of God. Any person vested with governmental or political authority who attributes hatred or violence to the will of God has desecrated the teaching and doctrines of my religion.

5) Scientists and theologians agree that freedom of speech and the rights of the accused are the best guardian for both science and religion. The First Amendment to the American Constitution, intended to protect freedom of speech in the United States of American, is cited above. For a more practical protection of science and religion, we need to include also the principles of the Fifth Amendment, because that Fifth Amendment offers protection from cruel and unusual punishment. In fact the Fifth Amendment is designed to prevent the practice common in theocratic societies of falsely accusing a

religious dissenter of a crime, and then isolating them in prison and torturing them both physically and emotionally.

<u>The Fifth Amendment to the Constitution of the United States of America</u>:
No person shall be held to answer for a capital, or otherwise infamous crime, unless on a presentiment or indictment of a Grand Jury, except in cases arising in the land or naval forces, or in the Militia, when in actual service or in time of War or public danger; ***nor shall any person be subject for the same offence to be twice put in jeopardy of life or limb; nor shall be compelled in any criminal case to be a witness against himself, nor be deprived of life, liberty, or property, without due process of law***; nor shall private property be taken for public use, without just compensation.

The defenders of western democracy like to point out that in a real democracy there are no "thought crimes," which means that no one can be charged with a crime because of what they think or believe. This crucial concept, that only an action can be a crime, is embodied mostly in the First and Fifth Amendments, although it may be implied or referenced in other parts of the American Constitution. This concept that only an action can be a crime, and that no one should be punished for their beliefs, is to be honored and celebrated by everyone who takes religion seriously, and who seriously wishes the same good things for others as for themselves. *Authoritarianism is the toxin strong enough to kill any nation.* And the road to authoritarianism usually begins with breaches in the freedoms protected by the First and Fifth amendments. Much of human history is the story of authoritarian states or empires that did their damage and then collapsed. The people then pick up the pieces and try again. But we do not repeat exactly the same patterns over and over again. We do make progress. Please consider the steps recommended here, intended to help us move toward the reconciliation of science and religion. We must move toward completion of this monumental but necessary task. The old truth about the destiny of our human community is as fresh and new as today's sunrise: If we do not hang together, we will each be hanged alone.

Again, in Matthew 5: 23-26, is God's call to worship, the call from Jesus inviting each of us to be reconciled to our brother, to our sister, to walk a mile in their shoes:

> 23. "Therefore, if thou art offering thy gift [to God] at the altar, and there rememberest that thy brother has anything against thee, 24. leave thy gift before the altar and go first to be reconciled to thy brother, and then come and offer thy gift. 25. Come to terms with thy opponent quickly while thou art with him on the way [to settle a dispute in court, to settle a dispute in life]; lest thy opponent deliver thee to the judge, and the judge to the officer, and thou be cast into prison. 26. Amen I say to thee, thou wilt not come out from it until thou hast paid the last penny."

And how can we be reconciled to our brother and sister if we discredit and disrespect their religious beliefs so greatly that we do not even ask what they are? How can we be reconciled to our fellow human beings if we do not know them, and how can we understand them if we do not know their religion? My position should be clear by now. My understanding of Jesus' teaching is that in order to be reconciled to my fellow human being *I must learn about their religion.* To make a genuine effort to learn about the religions of other people is not optional for a Christian. It is an obligation at least as great as the obligation to share the story of Jesus and of his teaching. And this is the teaching of Jesus when advising us to be reconciled to our brethren. In my home, a picture hangs on our wall that is a gift to me from my wife Catherine. It is a portrait of a Native American chief. He is strong, brave, serene. He looks outward and slightly upward, contemplating the world and taking it into his courageous heart. Printed beneath the portrait of this elder are these words: "All honorable men belong to the same tribe." The truth is simple. Knock, and it shall be opened to you. Even now, as you imagine yourself knocking on the door of enlightenment, the spirit of the Good Steward has already invited you in. You may ask why is it not also written "All honorable women belong to the same tribe." Women also need to be reminded of this noble truth. It is a quirk of the English language that the word "men" supposedly has a connotation of including both the males and females of the human species. But the actual exclusion of women from the institutions of law, decision-making and the exercise of government power is still visible. This is an ongoing sign of the problem that men (male men) bring to religion, the problem of making God a violent and sadistic warrior, an invented dramatic character who is contrary to Jesus and his merciful Father in Heaven. The violent warrior God, the God who combats bad behavior with violence and destruction is human, not divine. The Father in Heaven described by Jesus is more like a Mother in Heaven. She will not hurt her child. She hates to see her child suffer. The primacy of stewardship, the teachings of Jesus and the Buddha and of the Celts and the Native Americans speak to me, and what they say is: "We all belong to the same tribe, every day, including on those days when we are honorable as well as on those days when we are less than honorable."

Chapter Thirteen: The Old and New in the Kingdom of Heaven

Jesus said that the kingdom of heaven is like a householder who brings forth things both old and new from his storeroom (Matthew 13:51-52). He also said that he did not come to destroy the [old] law but to uphold it and add to it (Matthew 5: 17-20). Let's take a look.

The Ten Commandments as taught to me by the Roman church:

1) I am the Lord thy God, thou shalt not have false gods before me.

2) Thou shalt not take the name of the Lord thy God in vain. [No false oaths.]

3) Remember, keep holy the Sabbath.

4) Honor thy mother and thy father.

5) Thou shalt not kill.

6) Thou shalt not commit adultery.

7) Thou shalt not steal.

8) Thou shalt not bear false witness against thy neighbor.

9) Thou shalt not covet [jealously desire] thy neighbor's property.

10) Thou shalt not covet [jealously desire] thy neighbor's spouse.

Notice that these ten "commandments" are not really commandments. They are *prohibitions*. They give us a set of restrictions telling us what *not to do*. Only three and four are positive commandments: observe the sacred time of the Sabbath, and honor your parents.

Note also that to "honor" one's parents is not precisely the same thing as to "obey" one's parents, and not even the same thing as possessing exactly the same faith as one's parents. We honor our parents by being the best person we can be. We honor our parents by being good citizens, by being a person of integrity, a good steward, by striving to achieve in accordance with our ability. We may even need to correct errors made by our parents, change the view of reality that was taught to us by our parents in order to truly honor them.

The laws given to us in the Old Testament are like the rules that parents make for infants and young children. They tell us what kinds of behavior to avoid so that we will not be self destructive. These prohibitions do not necessarily spell out for us what we need *to do* in order to be truly good, in order to be more than just a person who is "not bad" or not evil.

What must we do to be good, to be worthy of membership in the kingdom of heaven? Jesus added in the New Testament guidance that is like the advice given by parents to adolescents and young adults. By means of the new law in the New Testament Jesus added the requirements of good stewardship. All that Jesus said regarding what we must do to be worthy of the kingdom of heaven is comprised of positive guidance. With Jesus' teaching, our moral expectations have been expanded. Our goal is no longer limited to avoidance of bad behavior, but to also do those things that make a person, and a community, truly good. This is the essence of the new "thing" brought forth from the storehouse of the kingdom of heaven, all that which tells us the meaning and signs of good stewardship, all that which tells us that all people are evaluated according to their conduct and not according to their ethnic heritage or their membership in a particular tribe or clan or church or nation.

The Beatitudes, the Standard of Evaluation in Matthew 25, the obligation to be reconciled to our neighbors before we can be reconciled to our God, all of these new things are added to the old to create what theologians have called a "new covenant" or new agreement between heaven and the people of Earth. If we comply with this guidance, we will be worthy of the kingdom of heaven, we will be worthy of keeping the planet Earth in our care and will most likely have even more placed in our care. This is not a human parent or warlord passing out punishments and rewards. This is simply Nature doing what it does in order to survive. It is neither more nor less than the rules by which the universe operates, the way eternal that has no name.

In my country, the United States of America, there are fundamentalists who argue that the Ten Commandments should appear on the outside of every court building. That would make some sense if we were living three thousand years ago and we were all conservative Jews. But, if we are living in today, and we see ourselves as a nation built upon Christian morality, and we want the fundamental moral law to be written on our courts of justice, then what should appear outside our court buildings is the standard of evaluation from Matthew 25:

> I was hungry and you gave me to eat; I was thirsty and you gave me to drink; I was a stranger and you took me in; naked and you covered me; sick and you visited me; I was in prison and you came to me.

because this is the social justice taught to us by the Gospel of Jesus Christ. This is the meaning of moral behavior and we will not survive as a civilization of intelligent beings unless we make the important distinction between social justice and punishment for offensive behaviors. No decent religion can be based upon fear, anger and acts of punishment. That is not religion. Please look at Jesus again. Does he appear to you to be motivated by fear and anger? I see in Jesus a man motivated by a profound understanding of human nature and of the human identity and potential, the potential to rise above destructive emotions and practice a real religion based upon stewardship.

The new law is not disconnected from the old, not in opposition to the old, but added to the old, a step forward in the development of an intelligent species. This is scientific information. When our species was in its infancy, we received the Ten Commandments. When we were in our emotional adolescence, full of vitality and potential, ready to conquer the world and do great things, we received the Beatitudes and the Standard of Evaluation, the cold hard truth that we will lose all that we have if we do not understand our obligation to be the caretakers of life on Earth.

Chapter Fourteen: The Origins of Morality

Moral teachers who are moral failures:
A pattern of behavior that is familiar to all Americans is the charismatic evangelist who inspires and influences the religious beliefs and practices of thousands, perhaps millions of people who attend a church and support a Christian denomination with monetary contributions. Then, someone can no longer cover the secret that this evangelist has committed and is probably continuing to commit the sexual offenses that he so vigorously denounces, and or financial fraud. He is soon found guilty. Perhaps he confesses and apologizes publicly, on the way to an extended vacation with his mistress in his mansion by the sea where his yacht is docked. Clearly, this is hypocrisy.

Another form of hypocrisy among the "religious" that has come to our attention is the reality that many Roman Catholic priests have been exposed as pedophiles who preyed on altar boys and choir boys in their parish churches. These boys were and are the children of parents who actively and wholeheartedly supported what they thought was the moral mission of the Catholic Church. They contributed money to the church, money that they had earned by the work of their own hands. So, where does this leave us? What is the problem here? Is it just that some men are hypocrites, and some of them become priests, pastors, ministers, evangelists? Or, is there some more meaningful explanation? How can it be that some of those who seek and rise to the highest positions of moral guidance in our society turn out to be deceivers and monumental moral failures? I believe there is a scientific explanation.

The Origin of Morals
The reason we find hypocrisy everywhere, in people from all walks of life, including in our religious institutions, is because of an error in our understanding of what makes a person moral. We have in our civilization the widespread and deeply held conviction that churches and religious

institutions teach us how to be moral. This is the error in our understanding of human nature. As soon as this error is corrected, the truth becomes awesomely clear; the mind breathes a deep refreshing breath of relief. We do not get our morality from any church or religious institution. We possess morality because our individual sense of right relationships, justice and morality all originate in that part of our brains that gives each of us a sense of *proportion*. This means two very important things. First, the mental ability that enables us to count and measure, and perform all of the ordinary functions of arithmetic and mathematics, including higher mathematics, is *proportion*; and this same sense of proportion gives us our perceptions of justice and what is morally right and morally wrong. Second, since we each have our own individual brains, we each develop our own individual sense of proportion and sense of morality. We can be categorized and grouped of course, such as those against the death penalty and those for it, those against abortion and those against the birth of unwanted children, those who believe sexual love outside of marriage is not a great offense, and those who believe it is, and on and on. Although we can be grouped, and we are, and we can be labeled by labels such as "free thinker," "conservative," "fundamentalist," "Jesus freak," "liberal," "permissive," "immoral," "rigid," "tight ass," and so on, we are still individuals, and we do each possess our own definitions of moral justice, moral rightness, and our individual definition of God or of the origin of life.

In the teaching of Islam, there is a wonderfully clear separation of moral behavior into five categories.

1) Obligatory: things that a moral person must do, or certainly should do. (Fard)
2) Meritorious or recommended. (Mandub)
3) Permitted or tolerated. (Mubah)
4) Reprehended, considered wrong but not as wrong as the last category. (Makruh)

5) Forbidden, punishable: the worst of human behavior. (Haram)

In terms of authoritarianism in the Moslem religion, there are mullahs or imams and other institutional authorities who claim to identify fixed or traditional moral law and this fixed law is called Shariah. This is also known as the "pathway." In the rigid or fundamentalist version of Islam, all of the categories of behavior are strictly defined and monitored by the authorities. In a modern state with democracy and freedom of religion, but with a dominant Islamic tradition, each individual would have some freedom to define each of these categories for themselves. An example would be the nation of Turkey, where women are free to choose modern standards of dress or traditional Moslem restrictions. Men can seek to closely supervise their sisters and daughters, or they can encourage the females in their families to exercise the individual freedom and civil rights that are expected by citizens of a modern democracy. The most obvious exercise of freedom for a woman would be to pursue an education that enabled her to practice a profession and support herself as well as making a substantial contribution to the support of her family. In any case, thinking about all of human behavior, both moral and immoral behavior, as falling under these five categories is a helpful way to manage one's personal commitment to a standard of behavior. One can see that these five categories of behavior must depend upon a sense of what is fair and what is unfair, what are the various patterns of behavior through which we meet our obligations to ourselves, to our neighbors and to our God.

What is the Sense of Proportion?
There is reason to believe, both from natural history and modern studies of the human brain, that our entire ability to count and measure, and calculate in all the ways that we can discover or invent in the field of mathematics, arises from a small center in the brain that provides us with a sense of proportion. This sense of proportion means a highly precise ability to

distinguish the size of one thing when compared to the size of another. This unique genetic capacity may in fact arise from some form of "number line" or line length in the brain. Two publications that I have read have helped me to form my viewpoint about measurement, counting and proportion.

The first is *The Science of Measurement: a historical survey*, by Herbert A. Klein (New York: Dover Publications, Inc., 1974). This book conveyed to me, as I am certain it would convey to any reasonable person, the crucial reality that there are no absolute values with regard to any quantitative or qualitative "thing" that we human beings care to measure. All measurements are comparisons and all measurements are approximations. This is true for length, weight, volume, time, and all things that a human being can conceive of that we might want to measure. This book also played a role, along with many others, in my personal journey toward the conclusion that all of mathematics is only the mental extension of *measuring and counting*, and nothing more. Mathematicians often assign far deeper and far more comprehensive attributes to mathematics. They often claim that the universe is mathematical. They claim that mathematics is logic and logic is mathematics. To me, none of these claims are valid. I believe that the only thing a mathematician has that others don't have, is a kind of agility in mental computation that is similar to the physical agility of a gymnast. That is why mathematicians as a group usually claim that any mathematician, however great they may be, will invariably do their best work before they are thirty years old. This age limitation is probably not absolutely true, but it is a good piece of evidence that mathematical agility is similar to the physical agility of gymnastics. For me, it is correct to think of mathematics as "mental gymnastics," and that is why some individuals have more skill than others. However, I do not believe that mathematicians are inherently geniuses or even more rational than others. They probably engage in irrational behavior the same as others. However, I do believe that mathematicians have a highly developed and very precise sense of proportion. There is a way that my theory can be tested. Since I argue that the sense of proportion is the source of our individual sense of morality, try the following as a sociological study. For two samples, one the general population and the other sample made up only of professional mathematicians. For this study, I would not include accountants as belonging to the mathematician sample, because the profession of accounting attracts some unethical people whose primary interest is the world of money, investment and tax avoidance. I would include in the mathematician sample engineers whose predominant duties include the duty to correctly apply mathematical measurements and calculations to determine the types and sizes of materials to be used in construction. Now, having "gathered" these two samples, for each sample, look at the crime rate. Surprised? I predict that the crime rate among professional mathematicians is practically zero. Don't you want to know why? Here we are thinking that all mathematicians are scientific geniuses. I disagree with that viewpoint. However, because of my conviction that mathematical ability grows out of a strong sense of proportion, I actually believe that mathematicians are not scientific geniuses, but they may be moral geniuses. They do not commit crimes and they do not engage in immoral behavior nearly as much as the general population. How could that be? Easy, by an equation: proportion equals morality.

The second publication is the article "A Head for Numbers," *Discover Magazine: the world of science*, July 1997, pp. 108-115, by Robert Kunzig. In this brief story, Mr. Kunzig describes brain research that supports the conclusion that we possess a neurological organ that functions as a "number line," which we use to compare the size of one thing with another. If this is true, as I believe it is, and if this "number line" or measuring guide is highly precise, then it is in fact the explanation as to why we humans are able to count and calculate and lesser animals cannot. A wide variety of information about brain development that I have incorporated over the years suggests to me that the reason we appear to be "divided" from the lower animals in technological ability is because we are able to conceive of "number" as an independent attribute of any material object or set of objects. Certain higher animals, such as dolphins, apes, dogs and birds possess sophisticated language skills, but they do not possess technology to a level that even begins to approach the human capacity for technology. I propose that the explanation for this is that measuring and counting is the primary technology that precedes all others. In order to create any functional artifact with our skilled hands, we first measure, and measuring always requires and includes counting (the units of measurement). Squirrels and raccoons have skilled hands. Some birds weave admirable works of art with their beaks and claws. But none of these animals can plan and then build a chair. Why? They appear to have the physical dexterity. *The reason is because their brains do not enable them to conceive of number as an independent attribute.* Even when we train and test chimpanzees or dolphins to count bananas or fish, I believe that what occurs in their brains is that they form concepts of the objects where "number" is part of the whole but only as an attribute incorporated into and attached to the whole. To clarify, *we* learn that four bananas is comprised of two separate independent concepts, "four" and "banana." And because we conceive of number as an independent attribute, it is obvious to us that we can have four cars, four dollars, four years, four inches, four miles, or four sentences. But for our awesome and respectable friends, both in the water and in the forest, "four bananas" is a single concept, and "four fish" is a single concept, and neither the dolphin nor the chimpanzee, nor any other animal, can separate out the "four" from the bananas or the fish to which the four is attached. So, for our animal friends, "three fish" and "two bananas" are separate concepts, different from "four fish" and "four bananas," but distinguishable. The one and only thing we can do that they cannot, and that makes us look like masters of the universe, is that we can separate the "number" from one object, and attach the number to any other object as appropriate. Therefore, my viewpoint is that our technological capacity grows out of this two-part skill: we can perceive and compare the size of one thing to another with great precision (number); and we can separate that proportional size (number) from any and all other attributes of an object or set of objects. This is the essence of our sense of proportion. But we don't use our sense of proportion only to make physical artifacts. We also use it to make social artifacts, such as morals, laws and systems of justice. We can even separate our justice into categories: social justice, legal justice, sexual justice, marital justice, racial justice, political justice, and just about any kind of justice one can imagine. Those who bet on horse races and gangs of criminals are found to have a code of conduct, a code of justice. Even though they might kill you, they will only kill you in accordance with their rules. And if you do not deserve to be killed according to their rules, you are as safe as if you were praying in church. Maybe safer.

My perception of the human identity is that we are a "calculating, learning observer." We do all three of these things, with great skill and aptitude for focus and concentration. When we discover a similar intelligence in other animals, such as primates or dolphins or dogs, what we really have is a form of "learning observer" only. Our student animals can observe and learn, but what we can do that they cannot do is calculate (count and measure). Although we marvel at what a chimpanzee or a dolphin can learn, they will never be taught to measure a board with a ruler, and cut it to a desired size to fit as part of an assembly they conceived and planned. They can observe. They can learn. They cannot calculate. And because they cannot calculate, they are unable to conceive of all the possibilities that counting and measuring and calculating encompass. The few genes that separate us from our primate ancestors are the genes that give us the capacity to calculate, for good or evil.

Were We Taught Morals?

Let's take a closer look at this argument. Let's look at ourselves, modern humans, and how it appears that a child is taught to be moral and what moral justice is and what legal justice is. And then lets look at primitive societies and how they put their morality and justice into practice.

First, how did I learn to be moral? That is, how do we learn to be moral? In 1988 Robert Fulghum had his book published, entitled *All I Really Needed to Know I Learned in Kindergarten* (Villard Books, Random House). This book was immediately popular and recognized as a piece of conventional wisdom about American society and civilization in general. The book's focused theme is that we learn fundamental behaviors of fairness, courtesy and proper social interactions very early in life. In fact, the doctorates who study human brain development and how we learn would argue that we are ready to learn about fairness from day one following birth. Remember, the human personality is social. Although we like to distinguish ourselves from ants, who have division of labor, and from bison, who roam the plains in herds, we too live in communities. We like to live in communities; we feel safer living in communities; and we have ten thousand years of history that tells us it is not normal for a single human being to live alone, although that does occasionally occur. In modern times, even individuals who live in their own private dwellings "alone," are far from being hermits. They are connected to and are an integral part of a human community. The social skills that enable us to live successfully in the community setting come naturally to us. When young children lack these common social skills, we consider them to be neglected, deprived, impaired, or "damaged." It is true that any individual can learn to be "anti-social." However, our brains make us ready, even as infants, to learn how to be with others, how to relate to others, what kinds of behavior will be considered to be "proper" and "fair" and what kinds of behavior will be considered "wrong" and "unfair."

Let me be more specific and recount some of my early childhood memories of learning on the street, learning what is fair and what is unfair. First, I remember an "aphorism" that was just as powerful as any of the Ten

Commandments in my childhood neighborhood: "Pick on somebody your own size!" It is so obvious what this means. It means that a large boy -- or girl -- should not be fighting physically with a small boy or girl. This was especially true if the age gap was two years or more. We all knew it was wrong to be a bully, wrong to match yourself with someone who was obviously smaller, younger, or weaker than you. We fought, but there were implied rules for "fighting fair." We fought, but like I once heard a pastor say, "We need to know how to fight," even when members of a church are fighting over something. This moral concept of being fairly matched for a fight, or contest, is deeply imprinted on human society and upon the American way of life. Sports are an important part of American life, and we are all familiar with the concept of "league" or level of skill that pervades the world of sports. Simply put, we would never consider it fair, or interesting to have a high school Junior Varsity football team play against a professional team from the National Football League. This concept of being properly matched, fairly matched, pervades all team sports and individual sports. It is perhaps most obvious in the most violent of all sports, boxing. Boxing is divided not by age or height but by body weight. Imagine that! We read about the importance of weights and measures in the Old Testament and the New Testament, and we can read the stamp on a gasoline pump that tells us that the state authority on weights and measures has tested and approved the pump as a device that accomplishes a fair "liquid volume" measurement for the sale of gasoline fuel. Weights and measures, proportion, fairness, size is fairness, proportion is fairness. On the street, in school, in commerce, in sports, proportion is justice. Fight fair. Play fair. I vividly recall my childhood peers calling out "No kicking!" Or no pulling hair. The rules were not necessarily rigid, and the two combatants could make rules for themselves, on an equal rights basis. Play fair was also a deeply held conviction in my childhood neighborhood. I recall fondly those summer days when we played stick ball in the street and we started by "choosing teams." A popular and respected boy would be spontaneously nominated to be "captain," or a boy confident of his social status or skill status in the group would say "I'll be captain," or "I want to be captain." The two captains would throw their fingers out in the "even-odds" finger game or "catch the bat" to determine which captain would get to choose first for the second player on his team, and then the third, and so forth. It is as clear to me as the sun shining on a friend's face, that it was a source of pride to be chosen quickly, and a source of disappointment, even mild shame, to be the one who was deemed slow or "spastic" and therefore chosen last. I also remember team adjustments being made toward the end of the choosing, so that no one would be left out and to avoid having the teams be totally mismatched with one clearly being able to "slaughter" the other. If one team was heavy with heavy hitters, they would get one or more of the "non-athletic" players to "even up" the team strengths. We knew what was fair, and I question whether we knew what was fair because we went to church. After a lifetime of experience and study, I believe that our sense of proportion is the true origin of our sense of morality and justice. There is one more reason why I take this position, and that is because human society existed for thousands of years without churches, and human society had to have a moral ground, had to have moral rules and a sense of fairness in order to survive. Our human society had this sense of morality millennia before we had churches.

Second, how does a primitive society render justice? There are a multitude of books and works in the field of anthropology, and numerous documentaries that tend to redeem the vast wasteland of modern television programming by providing us with interesting factual information about how primitive societies operate. From the wilds of Borneo to the jungles of Africa and South America, to isolated deserts and mountains, we find that primitive tribes usually employ systems of "restorative justice" and or "retributive justice." The concept of restorative justice is instantly familiar to us, because our Old Testament makes several references to "an eye for an eye and a tooth for a tooth." This concept of compensating the victim of an injustice with payment of exactly equal value is both harsh and starkly fair. We believe it and disbelieve it at the same time, because we have another extremely important concept of fairness and social justice that "two wrongs don't make a right." And so we see, as we begin to explore morality and justice on the most primitive levels, among children and among isolated, primitive tribes, we run into complications. Fairness and justice are not simple. Decisions become difficult. We do not feel right causing a second injury as the proposed means to remedy a first injury. Although restorative justice feels right to us it is still a profound challenge. How do you "restore" previous relationships, the previous state of the community, when one member has murdered another? Restoration seems so much simpler when the offense, the unfair behavior, was theft or vandalism. But how do you restore what an individual and a family has lost when a man unknown to the family has raped a sixteen-year-old girl who was innocent and clean of heart and who loved life and learning? Restorative justice is not simple, it is certainly not as simple as "an eye for an eye and a tooth for a tooth." Most people do not believe that executing a person who committed murder brings back the one who was murdered. Still, restorative justice is an exciting and promising field of human exploration. There are books and documentary films, a couple of which I have seen, which describe how even a murderer can be confronted with the family of his murder victim and made to experience the horror of what he has done. He can apologize, sincerely. The family can get some genuine satisfaction from conveying the depth of their loss directly to the offender -- in a safe environment of course. The offender, though not relieved of his sentence of imprisonment, can be enabled to do community service, participate in therapy, address the scope and impact of his offense and become far better prepared for genuine rehabilitation than if he had participated in no attempt, however monumental this may seem, to "restore" the social relationships of the community that existed before the murder and were broken by it.

I recall a documentary on television where a woman of one tribal group was murdered by another tribal group and the offended tribal leaders engaged in a discussion where they were saying "They will have to give us a woman." How childlike and innocent, yet how down-to-earth and positively fair this idea sounded. They did not want to "remedy" the loss by committing another killing, but by presenting to the other group the simple concept of moral fairness that you must match our loss with a loss of your own. We lost a woman; now you must lose a woman. And you must let us chose the woman that you lose. You cannot compensate us for our loss by sending us a woman that is disabled and a burden. We lost a young, attractive, skilled and strong woman. You must give us a young, attractive, skilled and strong woman in return. Whether or not this appears "primitive," even though there is in the rain forest no court building, no judge in robes, no paid attorneys, the outcome is similar: restorative justice is negotiated. This "system" is in fact, some would argue, more humane and more effective than ours. Our buildings and our robes, our paid attorneys and our tons of books on written laws and procedures, our thousands of years of experience, have not added any value to our justice. We still are limited in our ability to render justice for all by our simple inborn sense of fairness, our sense of proportion. Take the reality of how justice and morality appears to be neither more nor less than our innate sense of proportion, and place this body of experience next to our sad but devastatingly true experience of the moral corruption of religious institutions and the members of "holy orders," the centuries of authoritarianism and the execution of dissenters and "heretics," the burning of books, the banning of books, the sexual misbehavior of priests and the persistent internal conspiracy to protect the offenders rather than acknowledge the victims, and one must see that it is a hard argument to contend that we learn morality in church.

People leave churches because they hunger and thirst for justice. Please note that there is nothing in the New Testament, not a single word or act that tells us Jesus recommended we learn how to be good by entering into holy orders or by becoming a member of a religious institution. Jesus taught and practiced his morality on the street, in the fields, on the shores and in the boats of fishermen, at the table of the poor and distraught and the sick and sinners. The morality of Jesus does not grow out of a Christian church or a "Jesus" church or any kind of church. The morality of Jesus grows out of the obligations of the individual to the community and the reciprocal obligations of the community to the individual, the social contract, the sense of proportion and fairness in human relationships. If anyone doubts this assertion, please look again at the New Testament. Look at what Jesus says about human relationships. Look at, and count, his references to weights and measures, the measure of justice and fairness in terms of money, or coins, or commodities that are traded. See him throw the money-changers out of the Temple, because they were not there to teach morality. They were there for their own profit. Jesus rejects the self-serving authority of the Temple, because it is political, because it is used to keep selfish people in power. Morality is for Jesus, as it has been for humankind since the dawn of history, the sense of fairness that we feel in our heart, the voice of justice that speaks to us from the sense of proportion that is at the center of the human mind that makes us members of a community that not only longs for justice, but survives and thrives in accordance with its ability to be moral and just and fair. No matter what we do, no matter whether we are building bridges or rockets or fighting wars, we are still children on the street expecting and demanding fairness in human relations. Even when at war, the Pentagon, the most respected military institution in the world, has acknowledged the reality of "asymmetrical warfare," which is just a politically correct word for "unfair war." What could "asymmetrical warfare" mean except a war where the sides are not evenly matched, not fairly matched. The most technological advanced armed force ever on earth, with electronic sensing and guidance systems, against men who have only rifles and Improvised Explosive Devices and probably no uniform and maybe no shoes. The reality of proportion penetrates us. It is everywhere that human beings engage in any action, however social or anti-social it may be. Wherever we hear, see or taste, wherever we feel, think or speak, our sense of proportion is turned on, like a television or a radio that is never shut off. We are listening, watching, always, for what is fair, what is proper, what is just, what is moral. It is our sense of proportion that provides us with the answers, however simple or complex they may be. This is scientific information. This is consistent with the teaching and life of Jesus. Look again if you need to. See if you can find the place where Jesus said, "Blessed are those who attend church."

Chapter Fifteen: Stewardship is the Action of Love

"How do I love thee? Let me count the ways!" says the poem by Elizabeth Barrett Browning, #43 in the collection entitled "Sonnets From the Portuguese". And that concept of many ways of loving is an interesting concept, a profound concept. We do love in many ways. How many? And are some ways better than others, or just different? And is there a form of love, a way of loving, that is appropriate for every occasion? Stories of the agony and ecstasy of "romantic" love, another name for mutual instinctive chemical sexual arousal, pelt us like the monsoon. Soap operas, epic novels, tabloid magazines shout out the gossip we must know about people of fame or of infamy, or of the unknown ordinary masses, who find their way into episodes of extreme happiness, or extreme tragedy, addiction, sickness, murder, suicidal gestures, all associated with the power of romantic (sexual) love. When there is a murder and a private investigator is hired, the usual first act of investigation is to ask the question "Was it for love or money?" That is the question that leads us to the suspects, those who loved too much, or loved too little, those who stood to gain by removing the lover, or the unloved. How cosmically significant that the power of love is comparable to the power of money. I thought love was pure and noble.

"All you need is love!" says the Beatles song, "It's easy." Is that true? Is it easy? Is it easy to love other people? To love other people who are different? Is it easier to love people who live far away than to love those who live very close? Who is easy to love? Who is hard to love? And what kind of love is this that the song is about, love that is easy, love that is all that we need? Is it true that if we just loved, everything would be just fine, peaceful, fair, calm, happy? Or, are there people who are having a bad day, or a bad year, or a bad life, who will not be able to respond to our love, not be soothed, not stop screaming and lashing out, not stop hating and killing? Maybe there is something more than love that we need.

Or, maybe there is a special kind of love that we need that is the kind of love that everyone, even the screaming killer, responds to. Is that the love that Jesus was talking about?

It is said that love is giving, not taking. Giving according to one's ability means more is expected from those to whom much is given.

> Proportional responsibility; knowing what is expected. Luke
> 12: 48:
> 48. Whereas he who did not know it [the will of the master],
> but did things deserving of stripes [lashes with a whip], will
> be beaten with a few. But of everyone to whom much has
> been given, much will be required; and of him to whom they
> have entrusted much, they will demand the more.

This passage means that those who have the gifts of ability, skill, a great capacity to learn and accomplish things, are those "to whom much has been given." And they are expected to exercise their gifts for the benefit of the community and the society. The fact that they may "give more" than others who have lesser gifts does not entitle them to great or "disproportionate" rewards (compensation and social status). Therefore, proportional justice means that those individuals who have more to give are not really "giving more" for the simple reason that they are giving what they are able to give, the same as those with less ability, in terms of their economic contribution to the community. Each is giving what they are able. To have much to give, and refuse to contribute according to one's ability, is obviously a proportional injustice. This is like a successful fisherman saying: "Because my neighbor caught only a few, small fish today, I will do the same. Why should I 'give more' than he does?" In the story of the "Widow's Mite," in Mark 12: 41-44, Jesus observes a poor widow give two small coins for charity, and he tells his disciples that she has given more than all the rich, even those who have given thousands of denarii. How can that be, when she has given so little? Because charity and philanthropy are proportional to what one has. One's generosity is not measured by the amount one gives, but by the amount that one holds back to keep for oneself.

These passages also mean that if a greater responsibility is given to a person because they are judged able to manage greater responsibility, they will be held accountable *in proportion to* their ability and knowledge. This is a form of proportional justice that is not practiced in American society today. If it were, those who commit "white collar" crimes, college-educated people who have access to very high incomes, lawyers, accountants, and business advisors, but then lie, cheat and steal, and ruin the lives of employees by destroying their retirement funds, would be given greater consequences for their crimes than the consequences given for lesser crimes committed by those with lesser education, knowledge and ability, such as stealing a television.

What Jesus said here was interpreted by St. Paul the Evangelist and Karl Marx and applied to the concept of economic justice when they taught that a democratic, communist or "communalist Christian" society would apply the principle: "From each according to their abilities; to each according to their needs." This means that those who have much ability are expected to exercise their gifts for the benefit of society, and the fairness practiced in "proportional justice" means that their reward (compensation, social status) will not be greater than what they need. This is of course a bit strong. Most people want those who make great contributions to society to receive suitable rewards, but there is the *reasonable argument* that the proper reward for a great achievement is a social reward, such as recognition, respect, and the role of advisor, not more things and more money.

It is said that love means placing the needs of the other person ahead of your own. This is motherly love. It is said that the love of a mother for her child is "unconditional." She does not love the child and nurture and protect it because of some qualities or traits of the child. Because it is a child, the child, her child and our child, a child of the universe, and just because it exists as a child, in the sacred form of a child, in the sacred form of the continuation of the species, the mother loves the child and defends it with her life. Sick at night, tired, she will arise from her sleep to soothe the crying child. She will feed it before she feeds herself. She will take off her own clothes and expose herself to keep her child warm. This is the mythic but also real mother known here on Earth, human mothers and mothers of many animal species. Love is giving, love is sacrifice, love is capable of acts of great courage and selflessness, including the violence that might be necessary to protect the child from harm. In the real world of forest, jungle and plain, one is in great danger if perceived as a threat to a child. Mother is nearby, and you must go through her to get to the child. This is a form of love. This is scientific information. Is this the form of love we need? This is a kind of caring. This is care taking. Is this good stewardship? Is this easy? It doesn't look easy. I believe we are getting close. The unconditional love of the mother, the love that appears to be comprised of devotion and duty more than only self-interest, appears to be good stewardship and love at the same time. We are looking for the kind of love that is stewardship, and the kind of stewardship that is love. Jesus seemed to be telling us that all we need is a kind of love, but he also was most emphatically telling us that we must practice good stewardship in order to survive and thrive. There must be common ground then, a way of life where stewardship and love are one and the same act, both spiritually warm and coldly rational, loving as a duty and as a necessity, but still loving voluntarily, from a strong and willful heart. I perceive a clue, consistent with all of Jesus' teaching, in the parable of the good shepherd searching for his lost sheep. This story begins when Jesus is talking about children.

(Matthew 16: 10-14)

10. "See that you do not despise one of these little ones; for I tell you, their angels in heaven always behold the face of my Father in heaven. 11. For the Son of Man came to save what was lost. 12. What do you think? If a man have a hundred sheep, and one of them stray, will he not leave the ninety-nine in the mountains, and go in search of the one that has strayed? 13. And if he happen to find it, amen I say to you, he rejoices over it more than over the ninety-nine that did not go astray. 14. Even so, it is not the will of your Father in heaven that a single one of these little ones should perish."

The one vulnerable sheep is equal in importance to the ninety-nine that are safe. The sheep that is in danger needs to be brought back to safety. This is a pointer. It points to a pattern, a pattern that may be obvious but which I feel I still need to spell out. I have already cited the New Testament extensively. The pattern I see is the pattern of nurturing and liberation, nurturing those who are weak and vulnerable, liberating the strong. The pattern is obvious in the Sermon on the Mount (The Beatitudes) and in Matthew 25 (The Standard of Evaluation). The behavior where love and stewardship are the one and equal action is when we nurture the weak and liberate the strong. We could say, we should say, when we nurture *ourselves* when we are weak and liberate *ourselves* when we are strong. For anyone who has lived has experienced both. Jesus is telling us, all of us in our patriarchal societies, that we need to be more like the feminine mother. We need to be like the one who is known as the good mother and the good shepherd when she nurtures her child so long as the child is weak and vulnerable, but just as lovingly liberates the child, male or female, when that child is strong and ready to live life as a new and free member of the community. Many in the animal world nurture and protect their young ferociously, but then, when the child is strong, insist that they depart, go away and continue life without the protection and without any further guidance from the parent. It is time to go, time to learn on your own, time to fly. How familiar it is to us, in all cultures, that some little birds hesitate in fear and must be "pushed from the nest" so that they will learn how to fly. This is the love practiced by good human parents also. We recognize when the time has come to push our children outward, into the world in which they must live. Today in America, we are confronting this reality with new conditions, because our economy has made it harder for young adults to leave home. It is harder for them to earn enough money to pay for rent and transportation. It is harder for them to get a college education. If a child does get a college education, they begin their life of new-found "freedom" with a debt of tens of thousands of dollars in student loans. We have the common story, sometimes told as a joke, of the young man or woman who is twenty-five years old, or even thirty, who calls on the phone and hints that times are bad and maybe they could "come home" for a while. And we the parents, find the words, somewhat diplomatic words, suggesting that coming home is not a good idea.

Look at the New Testament again if you need to. Our repetitious Christ, teaching us, we who are in the Kindergarten of the universe, again and again that the key to survival of an intelligent species is this icon of *proportional justice* and *proportional love* -- nurture life when it is weak, liberate life when it is strong. The kind of love that we need is the good stewardship kind of love, the love that nurtures the weak and liberates the strong. This is not only the highest ideal of religion, it is also the highest ideal of government. The best government is not the government that governs least, but the government that enacts the good stewardship of the tribal or national community by liberating the strong and nurturing the weak so that they too will become strong again. This is the embodiment and implementation of the Golden Rule. Is this not how we all would like to be treated? To be healed and nurtured back to health when we are sick? To be given another chance when we have had a lapse of judgment and committed an offense? To have water when we thirst, food when we need nourishment? Company when we are lonely? Comfort when we are traveling away from home? It is so obvious, so clear that this is true religion, this is the way eternal that has no name, the way for an intelligent species to be what it was meant to be and do what it was meant to do.

And so, I say that when the Beatles sang "All you need is love," the song really means "All you need is good stewardship," because good stewardship, nurturing the weak and liberating the strong, is the kind of love that is all that we need. This form of love flows from a marriage of the warm passion of the spirit and the cold consideration of the mind. They who love best offer the best of what is best for the loved.

Listen to the song again. "All you need is love. Love is all you need." That second verse: "Love is all you need," is the important one. It means that when a living thing is receiving all that it needs, but not more than it needs, it is being loved. This is what "love" means. We know that a person or a thing is loved when it is receiving all that it needs. This is scientific information, about love, in the Gospels. Nurturing and liberation are the actions. We are each and all called upon to be the nurturer and the liberator, to love this way, the way of a Good Steward.

Chapter Sixteen: A Plan of Actions

This last chapter is the reader's chapter, a last word of encouragement to a traveler in this world. One can continue writing this book with one's life. Many laborers are needed and there is much that needs to be done. One's life is one's plan of actions, one's hopes and dreams of being nurtured and liberated. All we need is people who practice true religion and do not let their traditions get in the way. The act of love that is good stewardship is the heart and soul of true religion. All of the rituals and rules and doctrines established and defended by the World's Great Religions are not religions but rather the World's Great Traditions. We can honor our traditions, practice the old ways and genuinely enjoy them, but we should not allow any tradition to prevent us from practicing the true religion of good stewardship, which is not attached to any institution or organization but is just the way, the way of Christ, the way of the Buddha, the way of Moses in the desert and the way of Mohammed when he was not forced into the role of an aggressive warrior.

In the *Perennial Philosophy* (published in U.S. by Harper and Brothers, 1945), Aldous Huxley conveys the attitude of universal humanism. This wonderful book is Huxley's documentation of the observation that the human hunger for ultimate truth about who we are is what motivates the common holy spirit and common search for enlightenment, not the need for power and control or for limiting the freedom of others. The stewardship of our search for truth is the same stewardship that nurtures the weak and liberates the strong, for the greater good of the tribal or national community, as we are each and all together on the path to the ultimate truth that we seek.

The renowned Buddhist monk, Thich Nhat Hanh, advises us in his book *Living Buddha, Living Christ* (G. P. Putnam's Sons, 1995) that the wise course of action for each and all of us is to honor our traditions, but continue our journey to enlightenment. For those who experience a love for truth, a religious institution is not always the good servant and guide, but in fact a religious institution can be the obstacle to individual and societal enlightenment. In my inherited tradition, the tradition of the Roman Catholic Church, the authorities are fond of stating our "holy obligations" as in a certain holy day being a "holy day of obligation" meaning it is a day when one must attend church. There are so many other obligations. I recall that I was taught in my childhood that one of my obligations was to never enter a Protestant church, not to enter any non-Catholic church with a non-Catholic friend. Indeed, it was felt, though perhaps not stated explicitly, that having a non-Catholic friend was in itself morally dangerous. I am aware that Jews and Moslems sometimes give their children the same form of advice. This is the glaring defect in our traditions, the fact that they can be obstacles, often the greatest obstacle, to the kind of love that we all need. Therefore, I have cast off most if not all of the burdensome and unjust "obligations" of my tradition. Listening carefully to Jesus, I have accepted the one "holy obligation" that I believe is more holy than any other and more obligatory than any other, that is the obligation to be reconciled to my brethren before I bring my personal gifts to the altar of my God. I am trying to the best of my ability and that is all that Jesus asked. I cannot pretend that I do not understand what is required of me, because I do. The life and teaching of Jesus is not hidden, not concealed, not mysterious, not accessible only to experts. It is as clear as children playing in the sun on a summer's day here on Earth, the ultimate objects of our best love.

For love to be real, one must proceed from feeling and contemplation to action. I believe it was philosopher Edmund Burke who said: "All that is necessary for the triumph of evil is for good men [people] to do nothing." In my Christian tradition, I was taught as a child to pray. I recall a set of three prayers that were named the "Act of Faith," "Act of Hope," and "Act of Love." It is so very interesting that for a person to sit still, or lie still, or kneel and recite the words of a prayer, aloud or silently, is deemed to be "an act." For children and especially for adolescents the chemical hormones in the blood are overwhelming. Even in our early adult years, we experience love as a feeling, as a kind of force compelling us toward another, desire for closeness, intimacy, physical passion. As we grow older, and the body cools down, we see that love is not only what we feel, but what we do. Just as what we truly believe is revealed by what we do, how we love is revealed in what we do. An "act of love" is not a silent prayer. Religious traditions in all of the Great Religions press us inward to contemplation and silence and stillness, to soothe the tortured soul, the soul and the heart that is uncertain what to do. But the essence of what we need to do is not a quiet hidden mystery. We each need to find our way to our action of love, our action in the marketplace, on the street, in the community, not in our isolated distracted homes, but in the home of the society. One's action of love will be either nurturing or liberating, or both. One's action of good stewardship will be taking care, taking care of business or taking care of someone. It must be from your own plan or impulse, from your heart, and it must be something that you believe you do well. Your action of loving stewardship, whatever you choose, is you living in accordance with the Gospels. Your effort to take care of people or the things that people need, whether it be private or quiet, public or publicized, is the act of love that Jesus talked about, the act of nurturing and liberating ourselves and the world at the same time, with the same action. The Good Shepherd is not one who is climbing a mountain toward enlightenment. The Good Shepherd has visited enlightenment and has come back down to earth to be here among us who are busy, sweating, making a living. One's action of love occurs at a time that is inconvenient, at a place that is uncomfortable, under circumstances that are awkward, surrounded by chaos, uncertainty and fear.

Selected Bibliography for *The Primacy of Stewardship*

History

"Alexandria." *Encyclopedia Brittannica, Macropaedia.*
    (Vol. 13, pp. 238-241). Chicago: University of Chicago, 1991.

Baines, John and Jaromir Malek. *Ancient Egypt.* Alexandria (VA):
    Stonehenge Press, 1991.

Bauval, Robert. *The Orion Mystery: Unlocking the Secrets of the Pyramids.*
    New York: Crown Publishers, 1994.

Bennett, Ross, Editor, et al. *Lost Empires, Living Tribes.* Washington:
    National Geographic Society, 1982.

Brown, Dale M. et al. *Egypt: Land of the Pharaohs.* Alexandria:
    Time-Life Books, 1992.

Cahill, Robert E. *New England's Ancient Mysteries.* Salem: Old Saltbox
    Publishing, 1993.

Clark, R.T. Rundle. *Myth and Symbol in Ancient Egypt.* New York:
    Thames and Hudson, 1959.

Daniken, Erich von. *The Eyes of the Sphinx.* New York: Berkley, 1996.

David, A. Rosalie. *The Egyptian Kingdoms.* New York: E. P. Dutton and
    Elsevier Phaidon, 1975.

Davidson, David. *The Great Pyramid, Its Divine Message.* UK: Williams
    and Norgate, 1932.

Ellerbe, Helen. *The Dark Side of Christian History.* San Rafael:
    Morningstar Books, 1995.

Fell, Barry. *America B. C.* New York: Pocket Books, Simon and Schuster,
    1989.

Furneaux, Rupert. *Ancient Mysteries.* New York: Ballantine Books, 1977.

Gill, Joseph. *The Great Pyramid Speaks: an adventure in
    mathematical archaeology.* New York: Barnes and Noble, 1984.

Harpur, James. *The Atlas of Sacred Places.* New York: Henry Holt and
    Company, 1994.

International Edition. Encyclopedia Americana, (*Vol. 18, p. 432*). Danbury: Grolier, 1991.

Johnson, Edgar N. *An Introduction to the History of the Western Tradition*. Boston: Ginn, 1959.

Kielland, Else Christie. *Geometry in Egyptian Art*. UK: Alec Tiranti, 1959.

Knight, Christopher, and Robert Lomas. *The Hiram Key*. Rockport: Element, 1996.

Mackey, Albert G. *The History of Freemasonry*. New York: Gramercy Books, 1996.

Mendelsohn, Kurt. *The Riddle of the Pyramids*. UK: Thames and Hudson, 1975.

Murray, Margaret Alice. *Egyptian Sculpture*. New York: Charles Scribner's Sons, 1930.

Palmer, R. R., and Joel Colton. *A History of the Modern World*. New York: Knopf, 1962.

Peet, T. Eric. *The Rhind Mathematical Papyrus*. UK: Hodder and Stoughton, 1923.

Rice, Michael. *Egypt's Making: The Origins of Ancient Egypt*. New York: Routledge, 1990.

Smyth, Charles Piazzi. *Measurements of the Great Pyramid*. London: R. Banks, 1884.

Stecchini, Livio. "A History of Measures." *American Behavioral Scientist*. IV (1961) #7.

Stecchini, Livio. "The Origin of the Alphabet." *American Behavioral Scientist*. IV (1961) #6.

Tarhan, E. H. *Nur 4000 Kultur*. Ahlen, 1986.

Tompkins, Peter. *Secrets of the Great Pyramid*. New York: Harper and Row, 1971.

Vanderberg, Philipp, Thomas Weyr, translator. *The Curse of the Pharaohs*. Philadelphia: J. B. Lippincott Company, 1975. (Originally published in German, 1973.)

Velikovsky, Immanuel. *Worlds in Collision.* New York: Dell Publishing Co. Inc. , 1968.

Ward, Kaari, ed., et al. *Jesus and His Times.* Pleasantville: Reader's Digest Association, 1987.

West, John Anthony. *Serpent in the Sky.* New York: Julian Press, Crown Publishing Group, 1987.

Physics

Asimov, Isaac. *Understanding Physics.* (Three volumes in one text). New York: Barnes and Noble Books, 1993.

Barnes, Thomas. *Foundations of Electricity and Magnetism.* Boston: D. C. Heath and Company, 1965.

Calder, Nigel. *Einstein's Universe.* New York: Viking Press, 1979.

Cohen-Tannoudji, Gilles. *Universal Constants in Physics.* New York: McGraw Hill, Inc., 1993.

Coveney, Peter and Roger Highfield. *The Arrow of Time.* New York: Fawcett Columbine, 1990.

Davies, Paul. *Other Worlds.* New York: Simon and Schuster, 1980.

Davies, Paul. *The Mind of God.* New York: Simon and Schuster, Touchstone, 1993.

Einstein, Albert. *Relativity: the special and the general theory.* New York: Wings Books, Random House, 1961.

Fong, Peter. *Physical Science, Energy, and Our Environment.* New York: MacMillan Publishing Co., Inc., 1976.

Gardner, Martin. *Relativity Simply Explained.* Mineola: Dover Publications, Inc., 1997.

Gribbin, John. *Unveiling the Edge of Time*. New York: Harmony
  Books, 1992.

Hawking, Stephen. *A Brief History of Time*. New York: Bantam Books,
  1988.

Herbert, Nick. *Quantum Reality*. Garden City: Anchor Doubleday, 1985.

Jastrow, Robert. *God and the Astronomers*. New York: Warner Books,
  1978.

Lerner, Eric. *The Big Bang Never Happened*. London: Simon and
  Schuster, 1992.

Towne, Dudley. *Wave Phenomena*. New York: Dover Publications, 1967.

Will, Clifford. *Was Einstein Right?* New York: Basic Books, Inc., 1986.

Political Science

Byrns, Ralph and Gerald Stone. *Economics*. Glenview: Scott, Foresman
  and Company, 1981.

Cornford, Francis, translator. *The Republic of Plato*. New York: Oxford
  University Press, 1968.

Ebenstein, William. *Great Political Thinkers*. New York: Holt, Rinehart and
  Winston, 1960.

Griffith, Samuel, translator. *The Art of War, by Sun Tzu*. London: Oxford
  University Press, 1963.

Heilbroner, Robert. *The Worldly Philosophers*. New York: Simon and
  Schuster, 1986.

Knightley, Phillip. *The First Casualty*. New York: Harcourt, Brace,
  Jovanovich, 1975.

Levy, Leonard, Kenneth Karst and Dennis Mahoney, editors. *The First
  Amendment*. New York: MacMillan Publishing Company, 1990.
  (Selections from the Encyclopedia of the American Constitution.)

Machiavelli, Nicolo. *The Prince and the Discourses*. New York: Random
  House Modern Library, 1950.

Morris, Kenneth and Alan Siegel. *The Wall Street Journal Guide to Understanding Money and Investing*. New York: Fireside, Colophon, Simon and Schuster, 1993.

Steffens, Lincoln. *Autobiography of Lincoln Steffens*. New York: ----, 1931.

Stoessinger, John. *Why Nations Go To War*. New York: St. Martin's Press, 1993.

Psychology (and Sociology)

Anderson, Carol, Douglas Reiss, and Gerard Hogarty. *Schizophrenia and the Family*. New York: The Guilford Press, 1986.

Andrews, D. A., and James Bonta. *The Psychology of Criminal Conduct*. Cincinnati: Anderson Publishing Co., 1994.

Bass, Ellen and Laura Davis. *The Courage to Heal*. New York: Harper and Row, 1988.

Braun, Jay and Darwyn Linder. *Psychology Today, 4th edition*. New York: Random House, 1979.

Bredekamp, Sue, Editor. *Developmentally Appropriate Practice, birth to eight*. Washington: (NAEYC) National Association for the Education of Young Children, 1987.

Burgess, Ann, Nicholas Groth, Lynda Holmstrom, and Suzanne Sgroi. *Sexual Assault of Children and Adolescents*. Lexington: D. C. Heath and Company, 1978.

Butterfield, Fox. *All God's Children: the Bosket family and the American tradition of violence*. New York: Avon Books, 1996.

Dobzhansky, Theodosius. *The Biological Basis of Human Freedom*. New York: Columbia University Press, 1956.

Douglas, John et al. *Crime Classification Manual*. San Francisco: Jossey Bass Publishers, 1992.

Douglas, John and Mark Olshaker. *Journey into Darkness*. New York: Simon and Schuster, Pocket Books, 1997.

Douglas, John and Mark Olshaker. *Mind Hunter*. New York: Simon and Schuster, Pocket Books, 1995.

Dreikurs, Rudolph, M.D. *The Challenge of Child Training*. New York: Hawthorn Books, 1972.

Erikson, Erik. *Childhood and Society, 2nd edition*. New York: W. W. Norton and Co., Inc., 1963.

Erikson, Erik. *Gandhi's Truth*. New York: W. W. Norton and Co., Inc., 1963.

Fancher, Raymond. *Pioneers of Psychology, 2nd edition*. W. W. Norton and Co., Inc., 1990.

Frankl, Viktor. *Man's Search for Meaning*. New York: Pocket Books, Simon and Schuster, 1976.

Freud, Sigmund. *A General Introduction to Psychoanalysis*. New York: Washington Square Press, 1962.

Fromm, Erich. *The Art of Loving*. New York: Bantam Books, 1963.

Fromm, Erich. *Escape From Freedom*. New York: Avon, Holt, Rinehart and Winston, 1965.

Ginsburg, Herbert and Sylvia Opper. *Piaget's Theory of Intellectual Development*. Englewood Cliffs (NJ): Prentice-Hall, Inc., 1969.

Gladwell, Malcolm. "Damaged." *The New Yorker*, February 24 and March 3, 1997, pp. 132-147.

Glasser, William, M. D. *Reality Therapy*. New York: Harper and Row, 1965.

Grinder, John and Richard Bandler. *TRANCE-formations*. Moab (UT): Real People Press, 1981.

Herman, Judith, M. D. *Trauma and Recovery*. New York: Basic Books, Harper Collins, 1992.

Hook, Sidney, editor. *Determinism and Freedom*. New York: Collier Books, 1961.

Houston, John. *Fundamentals of Learning and Memory, third edition*. New York: Harcourt, Brace, Jovanovich, 1986.

Hunt, Morton. *The Story of Psychology*. New York: Doubleday, Anchor, 1993.

Hunt, Morton. *The Universe Within*. New York: Simon and Schuster, Touchstone, 1982.

James, Beverly. *Treating Traumatized Children*. New York: Free Press, Simon and Schuster, 1989.

Jourard, Sidney. *Disclosing Man to Himself*. Princeton (NJ): D. van Nostrand Co., Inc., 1968.

Jung, Carl. *Man and His Symbols*. New York: Doubleday and Company, Inc., 1964.

Kovel, Joel, M. D. *A Complete Guide to Therapy*. New York: Pantheon Books, 1976.

Laurence, Jean-Roch and Campbell Perry. *Hypnosis, Will and Memory: a psycho-legal history*. New York: The Guilford Press, Inc., 1988.

Lew, Mike. *Victims No Longer*. New York: Harper and Row, 1990.

Maslow, Abraham. *Toward a Psychology of Being*. New York: D. van Nostrand Company, 1968.

Minuchin, Salvador. *Families and Family Therapy*. Cambridge (MA): Harvard University Press, 1974.

Morgan, Elaine. *The Descent of Woman*. New York: Bantam, Stein and Day, 1972.

Nash, J. Madeleine. "Fertile Minds." *Time*, February 3, 1997, pp. 48-56.

Neill, A. S. *Summerhill: a radical approach to child rearing*. New York: Hart Publishing Co., 1960.

Orloff, Judith, M. D. *Second Sight*. New York: Warner Books, 1996.

Peck, M. Scott. *The Different Drum*. New York: Touchstone, Simon and Schuster, 1987.

Peck, M. Scott. *People of the Lie*. New York: Simon and Schuster, 1983.

Peck, M. Scott. *The Road Less Traveled*. New York: Simon and Schuster, 1978.

Reiser, Morton. *Memory in Mind and Brain*. San Francisco: Harper Collins Basic Books, 1990.

Restak, Richard. *The Brain*. New York: Bantam Books, 1984.

Sacks, Oliver. *The Man Who Mistook His Wife for a Hat*. New York: Simon and Schuster Summit Books, 1985.

Sagan, Carl. *Broca's Brain*. New York: Random House, 1979.

Sagan, Carl. *The Dragons of Eden*. New York: Ballantine Books, 1977.

Sass, Louis. *The Paradoxes of Delusion*. Ithaca (NY): Cornell University Press, 1994.

Satir, Virginia. *Conjoint Family Therapy, revised edition*. Palo Alto (CA): Science and Behavior Books, Inc., 1967.

Skinner, B. F. *Beyond Freedom and Dignity*. New York: Bantam, Alfred A. Knopf, Inc., 1971.

Skinner, B. F. *Walden Two*. New York: MacMillan Publishing Company, 1948.

Spitzer, Robert et al. *Diagnostic and Statistical Manual-III-R*. Washington: American Psychiatric Association, 1987. (There are some changes in subsequent editions, such as the DSM-IV-R ("Revised.")

Spretnak, Charlene. "Resurgence of the Real." *Utne Reader*, August, 1997, pp. 59-63, 106.

Sternberg, Robert and Kolligian, Jr. *Competence Considered*. New Haven: Yale University Press, 1990.

Szaz, Thomas. *The Myth of Mental Illness*. New York: Harper and Row, 1974.

Wadia-Ells, Susan, editor. *The Adoption Reader*. Seattle: Seal Press, 1995.

Yalom, Irvin. *The Theory and Practice of Group Psychotherapy, 2nd edition*. New York: Basic Books, 1975.

Religion

Bible: *Holy Bible.* O. T. Confraternity-Douay Version; N. T. Confraternity
Edition, Confraternity of Christian Doctrine. New York: Catholic
Book Publishing Company, 1957. (I believe this translation was the
most accurate ever written and will remain so until we obtain more
information about ancient languages. This evaluation is still subject
to the criticism that the church authoritarians removed extremely
important information that did not support their political and
doctrinal goals.)

Baigent, Michael, Richard Leigh and Henry Lincoln. *Holy Blood, Holy Grail.*
New York: Dell, 1983.

Berry, Ray. *The Spiritual Athlete.* Olema (CA): Joshua Press, 1992.

Bhaktivedanta, Prabhupada. *Bhagavad-Gita, as it is.* Los Angeles:
International Society for Krishna Consciousness, 1972.

Blakney, Raymond, translator. *The Way of Life, Lao Tzu.* New York: New
American Library, 1955.

Borg, Marcus. *Meeting Jesus Again for the First Time.* San Francisco:
Harper Collins, 1994.

Capra, Fritjof. *The Tao of Physics.* New York: Bantam Books, 1977.

Caussade, Jean Pierre de. Translated by Kitty Muggeridge. *The Sacrament
of the Present Moment.* San Francisco: Harper and Row, 1989.

Cotterell, Arthur. *MacMillan Illustrated Encyclopedia of Myths and
Legends.* New York: MacMillan Publishing, 1989.

Crossan, John Dominic. *Jesus: a Revolutionary Biography.* San Francisco:
Harper, 1994.

Dawood, N. J., translator. *The Koran.* New York: Penguin, 1977.

Eisenman, Robert and Michael Wise. *The Dead Sea Scrolls Uncovered.* New
York: Barnes and Noble, 1994.

Funk,Robert, et al. *The Five Gospels.* New York: Macmillan Publishing,
1993.

Gaer, Joseph. *What the Great Religions Believe*. New York: New American Library, 1963.

Garrison, Webb. *Strange Facts About the Bible*. Nashville: Festival Books, 1980.

Hamilton, Edith. *Mythology*. New York: New American Library, 1942.

Herbert, Edward, Editor. *A Confucian Notebook*. Rutland (VT): Charles E. Tuttle, 1992.

Heschel, Abraham Joshua. *The Sabbath*. New York: Farrar, Straus and Giroux, 1995.

Hughes, Philip. *A Popular History of the Catholic Church*. Garden City: Doubleday, Image, 1954.

Hummel, Charles. *The Galileo Connection*. Downer's Grove: Intervarsity Press, 1986.

Kittler, Glenn. *The Papal Princes*. New York: Dell Publishing, 1961.

Lewis, C. S. *Mere Christianity*. New York: MacMillan, Collier Books, 1960.

Lewy, Gunter. *The Catholic Church and Nazi Germany*. New York: McGraw-Hill Book Company, 1964.

Lincoln, Henry. *The Holy Place*. London: Corgi Transworld, 1991.

Magida, Arthur. *How to Be a Perfect Stranger*. Woodstock (VT): Jewish Lights Publishing, 1996.

Mahadevan, T. M. P. *Upanisads*. New Delhi: Arnold Heinemann Publishers, 1975.

Matthews, Caitlin. *Celtic Devotional, Daily Prayers and Blessings*. New York: Harmony Books, Crown Publishers, 1996.

Mollenkott, Virginia. *Godding: Human Responsibility and the Bible*. New York: Crossroad Publishing Co, 1987.

Moore, Thomas. *Care of the Soul*. New York: Harper Collins, 1992.

O'Murchu, Diarmuid. *Quantum Theology*. New York: The Crossroad Publishing Company, 1998.

Pandit, M. P. *The Upanishads*. Madras: Ganesh and Co., 1968.

Parrinder, Geoffrey, Editor. *World Religions: from ancient history to the present*. New York: Facts on File, 1983.

Potter, Dr. Charles. *The Lost Years of Jesus Revealed*. Greenwich (CT): Fawcett Publications, 1962.

Progoff, Ira. *The Cloud of Unknowing*. New York: Dell Laurel, 1983.

Prophet, Elizabeth Clare. *The Lost Years of Jesus*. Malibu (CA): Summit University Press, 1984.

Rahula, Walpola. *What the Buddha Taught*. New York: Grove Press, 1974.

Raymo, Chet. *Skeptics and True Believers*. New York: Walker and Company, 1998.

Rumi, Coleman Barks translator. *Feeling the Shoulder of the Lion*. Putney (VT): Threshold Books, 1991.

Schonfield, Hugh. *The Passover Plot*. New York: Bantam Books, 1965.

Severy, Merle, et al. *The World's Great Religions*. Pleasantville: Reader's Digest Association, 1978.

Shah, Idries. *The Sufis*. Garden City: Doubleday and Company, 1971.

Sohl, Robert and Audrey Carr, editors. *The Gospel According to Zen*. New York: New American Library, 1970.

Speight, R. Marston. *God is One: The Way of Islam*. New York: Friendship Press, 1989.

Spong, John Shelby. *Rescuing the Bible from Fundamentalism*. San Francisco: Harper, 1991.

Spong, John Shelby. *Why Christianity Must Change or Die*. San Francisco: Harper, 1998.

Thich Nat Hanh. *Living Buddha, Living Christ*. New York: G. P. Putnam's Sons, Riverhead Books, 1995.

Thomas, E. J., Translator. *The Perfection of Wisdom, selection of Mahayana Scriptures*. Rutland (VT): Charles E. Tuttle, 1992.

Unnamed editor. *The Wisdom of Moses Maimonides*. Mt. Vernon: Peter Pauper Press, 1963.

Wilber, Ken. *The Marriage of Sense and Soul: integrating science and religion*. New York: Random House, 1998.

Wilson, Ian. *The Shroud of Turin*. Garden City: Image, Doubleday and Company, 1979.

Yallop, David. *In God's Name*. New York: Bantam Books, 1984.

Yates, Gerard (S. J.). *Papal Thought on the State*. New York: Appleton-Century-Crofts, 1958.

Science and Popular Science

Alexander, Jane, editor. *1973 Nature-Science Annual*. New York: Time-Life Books, 1973.

Berlitz, Charles. *World of Strange Phenomena*. New York: Wynwood PressTM, 1988.

Calkins, Carroll, editor. *Mysteries of the Unexplained*. Pleasantville: Reader's Digest Association, 1982.

Colborn, Theo, Dianne Dumanoski and John Peterson Myers. *Our Stolen Future*. New York: Penguin Books, 1997.

Davis, Joel. *Journey to the Center of Our Galaxy*. Chicago: Contemporary Books, Inc.,   1985.

Field, Michael and Martin Golubitsky. *Symmetry in Chaos*. Oxford: Oxford University Press, 1995.

Fuller, John. *Arigo: Surgeon of the Rusty Knife*. New York: Pocket Books, Simon and Schuster, 1975.

Gleick, James. *Chaos: making a new science*. New York: Viking Penguin, Inc., 1987.

Greenfield, Susan, general editor. *The Human Mind Explained*. New York: Henry Holt and Company, 1996.

Hardwick, E. Russell. *Chemistry*. New York: Blaisdell Publishing Company (Ginn), 1965.

Hodge, Paul. *Galaxies*. Cambridge (MA): Harvard University Press, 1986.

Hoyle, Fred. *The Intelligent Universe.* New York: Holt, Rinehart and Winston, 1984.

Huxley, Julian. *Evolution in Action.* New York: Harper and Brothers, 1953.

Kern, Edward. "A Tree that Rewrote History." *Time-Life Books,* 1973 Nature-Science Annual, pp. 72-87.

Klein, Herbert Arthur. *The Science of Measurement: a historical survey.* New York: Dover   Publications, Inc., 1974.

Kolbert, Elizabeth. *Field Notes From a Catastrophe.* New York: Bloomsbury USA, 2006.

Kunzig, Robert. "A Head for Numbers." *Discover: the world of science,* July 1997, pp. 108-115.

Macrone, Michael. *Eureka!* New York: Harper Collins, 1994.

Morgan, Elaine. *The Aquatic Ape.* New York: Stein and Day, 1982.

Morehouse, David. *Psychic Warrior.* New York: St. Martin's Press, 1996.

Murchie, Guy. *The Seven Mysteries of Life.* Boston: Houghton-Mifflin Company, 1978.

Odum, Eugene. *Ecology and Our Endangered Life-Support Systems, 2nd edition.* Sunderland (MA): Sinauer Associates, Inc., 1993.

Ostrander, Sheila and Lynn Schroeder. *Psychic Discoveries Behind The Iron Curtain.* Englewood Cliffs (NJ): Prentice-Hall, Inc., 1970.

Schroeder, Gerald. *The Science of God.* New York: The Free Press, 1997.

Schwaller de Lubicz, R. A. Translated by Robert and Deborah Lawlor. *Symbol and the Symbolic: Egypt, Science and the Evolution of Consciousness.* Brookline (MA): Autumn Press, 1978.

Shapiro, Robert. *Origins.* New York: Summit Books, 1986.

Sobel, Dava. *Longitude.* New York: Penguin Books, 1996.

Tesla, Nikola and David Childress. *The Fantastic Inventions of Nikola Tesla.* Stelle (IL): Adventures Unlimited Press, 1993.

Tompkins, Peter and Christopher Bird. *The Secret Life of Plants.* New York: Avon Books, 1974.

Unnamed Editor. *Strange Stories, Amazing Facts*. London: Reader's
    Digest   Association, Ltd., 1975.

Whitten, Kenneth, Kenneth Gailley and Raymond Davis. *General
    Chemistry, Third Edition*. Philadelphia: Saunders College
    Publishing, 1988.

Zeki, Semir. *A Vision of the Brain*. Oxford: Blackwell Scientific
    Publications, 1993.

Superior Beings Visiting Earth

Bramley, William. *The Gods of Eden*. New York: Avon Books, 1990.

Bryan, C. D. B. *Close Encounters of the Fourth Kind: Alien Abduction,
    UFO's, and the Conference at M.I.T.* New York: Knopf, Inc., 1995.

Daniken, Erich von. *Chariots of the Gods?* New York: Bantam Books, 1974.

Daniken, Erich von. *Gods From Outer Space*. New York: G. P. Putnam's
    Sons, 1968.

Daniken, Erich von. *In Search of Ancient Gods*. New York: G. P. Putnam's
    Sons, 1974.

Darlington, David. *Area 51: The Dreamland Chronicles*. New York: Henry
    Holt and Company, Inc., 1997.

Good, Timothy. *Above Top Secret*. New York: William Morrow and
    Company, Inc., 1988.

Good, Timothy. *Alien Contact*. New York: William Morrow and Company,
    Inc., 1993.

Hall, Richard. *Uninvited Guests*. Santa Fe (NM): Aurora Press, 1988.

Jordan, Debbie and Kathy Mitchell. *Abducted!* New York: Carroll and Graf
    Publishers/Richard Gallen, 1994.

Keel, John. *Why UFO's?* New York: Manor Books, Inc., 1976.

Mack, John, M. D. *Abduction*. New York: Charles Scribner's Sons, 1994.

Randle, Kevin and Donald Schmitt. *UFO Crash at Roswell*. New York:
    Avon, 1991.

Randles, Jenny. *UFO's and How to See them*. New York: Barnes and Noble, 1997.

Randles, Jenny and Peter Hough. *Encyclopedia of the Unexplained*. New York: Barnes and Noble, 1995.

Strieber, Whitley. *Breakthrough*. New York: Harper Collins, 1995.

Strieber, Whitley. *Communion*. New York: Avon Books, 1988.

Strieber, Whitley. *Majestic*. New York: G. P. Putnam's Sons, 1989.

Strieber, Whitley. *Transformation*. New York: Avon Books, 1988.

Sullivan, Walter. *We Are Not Alone*. New York: McGraw-Hill Book Company, 1964.

Watts, Alan. *UFO Visitation: preparing for the twenty-first century*. London: Blandford, 1996.

Selections of "Pyramidology"

Bauval, Robert. *The Orion Mystery: Unlocking the Secrets of the Pyramids*. New York: Crown Publishers, 1994.

Clark, R.T. Rundle. *Myth and Symbol in Ancient Egypt*. New York: Thames and Hudson, 1959.

Daniken, Erich von. *The Eyes of the Sphinx*. New York: Berkley, 1996.

Davidson, David. *The Great Pyramid, Its Divine Message*. UK: Williams and Norgate, 1932.

Furneaux, Rupert. *Ancient Mysteries*. New York: Ballantine Books, 1977.

Gill, Joseph. *The Great Pyramid Speaks: an adventure in mathematical archaeology*. New York: Barnes and Noble, 1984.

Kielland, Else Christie. *Geometry in Egyptian Art*. UK: Alec Tiranti, 1959.

Knight, Christopher, and Robert Lomas. *The Hiram Key*. Rockport: Element, 1996.

Mackey, Albert G. *The History of Freemasonry*. New York: Gramercy Books, 1996.

Mendelsohn, Kurt. *The Riddle of the Pyramids*. UK: Thames and Hudson, 1975.

Peet, T. Eric. *The Rhind Mathematical Papyrus*. UK: Hodder and Stoughton, 1923.

Smyth, Charles Piazzi. *Measurements of the Great Pyramid*. London: R. Banks, 1884.

Stecchini, Livio. "A History of Measures." *American Behavioral Scientist*. IV (1961) #7.

Stecchini, Livio. "The Origin of the Alphabet." *American Behavioral Scientist*. IV (1961) #6.

Tarhan, E. H. *Nur 4000 Kultur*. Ahlen, 1986.

Tompkins, Peter. *Secrets of the Great Pyramid*. New York: Harper and Row, 1971.

West, John Anthony. *Serpent in the Sky*. New York: Julian Press, Crown Publishing Group, 1987.